CRUSOE'S ISLAND:

A

Ramble in the Footsteps of Alexander Selkirk.

WITH

SKETCHES OF ADVENTURE

IN

CALIFORNIA AND WASHOE.

BY J. ROSS BROWNE,

AUTHOR OF

"ETCHINGS OF A WHALING CRUISE," "YUSEF," &c.

NEW YORK:

HARPER & BROTHERS, PUBLISHERS,

FRANKLIN SQUARE.

1871.

By J. ROSS BROWNE.

CONTENTS.

CRUSOE'S ISLAND.

A DANGEROUS JOURNEY.

OBSERVATIONS IN OFFICE.

A PEEP AT WASHOE.

LIST OF ILLUSTRATIONS.

CRUSOE'S ISLAND.

A DANGEROUS JOURNEY.

OBSERVATIONS IN OFFICE.

A PEEP AT WASHOE.

CRUSOE'S ISLAND.

CHAPTER I.

THE BOAT ADVENTURE.

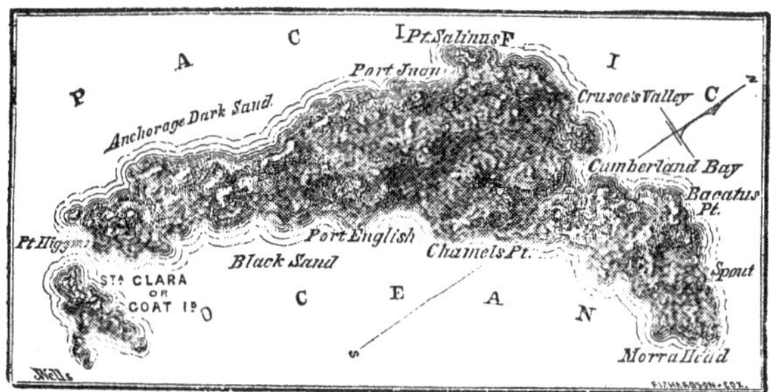

MAP OF JUAN FERNANDEZ.

My narrative dates as far back as the early part of the year 1849. Then the ship Anteus was a noted vessel. Many were the strange stories told of strife and discord between the captain and the passengers; pamphlets were published giving different versions of the facts, and some very curious questions of law were involved in the charges made by both parties. It appeared from the statement of the passengers, who were for the most part intelligent and respectable Americans, that, on the voyage of the Anteus to California, their treatment by the captain was cruel and oppressive in the extreme; that, before they were three weeks from port, he had reduced them almost to a state of absolute starvation; and, in consequence of the violence of his conduct, which, as they alleged, was without cause or provocation on their part, they considered their lives endangered, and resolved upon making an appeal for his removal at the port of Rio. On the arrival of the vessel at Rio the captain was arraigned before the American consul, and pronounced to be insane by the evidence of six physicians and by the testimony of a large majority of the passengers. It was charged, on the other hand, that the passengers were disorderly, mutinous, and ungovernable; that they had entered into a conspiracy against the captain, and in testifying to his insanity were guilty of perjury. The examination of the case occupied several weeks before the American consul; voluminous testimony was taken on both sides; the question was submitted to the American minister, to the British consul, and to the principal merchants of Rio, all of whom concurred in the opinion that, under the circumstances, there was but one proper course to pursue, which was, to remove the captain from the command of the vessel. He was accordingly deposed by the American consul, and a new captain placed in the command. This was regarded by the principal merchants of New York as an arbitrary exercise of authority, unwarranted by law or precedent, and a memorial was addressed by them to the President of the United States for the removal of the consul. A new administration had just come into power; and the consul was removed, ostensibly on the ground of the complaints made against him; but, inasmuch as some few other officers of the government were removed at the same time without such ground, it may be inferred that a difference in political opinion had some weight with the administration.

It is not my intention now to go into any argument in regard to the merits of this case; the time may come when justice will be done to the injured, and it remains for higher authority than myself to mete it out. I have simply to acknowledge, with a share of the odium resting upon me, that I was one of the rebellious passengers in the Anteus. My companions in trouble so far honored me with their confidence as to give me charge of the case. I was unlearned in law, yet possessed some experience in sea-life; and believing that the lives of all on board depended upon getting rid of a desperate and insane captain, aided to the best of my ability in having a new officer placed in the command. To the change thus made, unforeseen in its results, I owe my eventful visit to the island of Juan Fernandez.

It was the intention of our first captain to touch at Valparaiso for a supply of fresh provisions. In the ship's papers this was the only port designated on the Pacific side except San Francisco. Our new commander, Captain Brooks, assumed the responsibility of leaving the choice between Valparaiso and another port to the passengers. It was put to the vote, and decided that we should proceed to Callao, so that we might pass in sight of Juan Fernandez, and have an opportunity of visiting Lima, "the City of the Kings."

Early on the morning of the 19th of May, 1849, we made the highest peak of Massa Tierra, bearing N.N.W., distant seventy miles. The weather was mild and clear. As the sun rose, it fell calm, and the ship lay nearly motionless. A light blue spot, scarce bigger than a handspike, was all that appeared in the horizon. It might have passed for a cloud but for the distinctness of its outline. Weary of the gales we had encountered off Cape Horn, it was a pleasant thing to see a spot of earth once more, and there was not a soul on board but felt a desire to go ashore. For some days past, myself and a few others had talked secretly among ourselves about making the attempt in case we went close enough; but now there seemed to be every prospect of a long calm, and we took it for granted the captain would clap on all sail if we took the trades. There was no other chance but to lower one of the boats and row seventy miles. A party of us agreed to do this, provided we could get a boat. The ship's boats we knew it would be impossible to get without permission of the captain, and that we were not willing to ask. Mr. Brigham, a fellow-passenger, was owner of one of the quarter-boats. We broached the matter to him, and he gladly joined in the adventure, together with his partner and some friends, so that we made in all a very pleasant party of eleven. The proper number of men for the boat was six, but in consideration of the great distance and the necessity of a change at the oars, five more were crowded in. We had been in the habit of rowing about the vessel whenever it was calm, and this we thought would be a good excuse for lowering the boat. Being in great haste, lest the captain should object to letting us go, we only thought of a few necessary articles in case we should be cast away or driven off from the island. Two small demijohns of water, a few biscuits, a piece of dried beef, and some cheese and crackers comprised our entire stock of provisions; and for nautical instruments we had only a lantern and a small pocket compass. Not knowing but there might be outlaws or savages ashore who might undertake to murder us, we armed ourselves with a double-barreled gun, a fusee, and an old harpoon, which was all we could smuggle into the boat in the excitement of starting. Captain Brooks happening to come on deck, perceived that there was something unusual going on, and, suspecting our design, took occasion to warn us of the folly of such an expedition. At the same time, thinking there was more bravado than reality about it, he laughed good-humoredly when we acknowledged that we were going ashore. "Be sure," said he, as we went over the side, "not to forget the peaches. You will find plenty of them up in the valleys. Only don't lose sight of the vessel. You may exercise yourselves as much as you please, but keep the royals above water, whatever you do. Bear in mind that you are more than seventy miles from that peak!" We promised him that we would take care of ourselves, and come back safe in case we were not foundered.

At 9 A.M. we bade our friends good-by, and with three cheers pushed off from the ship. The boat was only twenty-two feet long and an eighth of an inch thick:

it was made of sheet-iron, and was very narrow and crank. Most of us, except myself and a whaleman named Paxton, were unused to rowing, so that the prospect of reaching land depended a good deal upon the day remaining calm, and upon keeping the boat trimmed, the gunwales being only ten inches out of the water.

LEAVING THE SHIP.

There was no excuse for this risk of life, save that insatiable thirst for novelty which all experience to some extent after the monotony of a long voyage. I will only say, in regard to myself, that I was too full of joy at the idea of a ramble in the footsteps of Robinson Crusoe to think of risk at all. If there was danger, it merely served to give zest to the adventure.

By a calculation of the distance and our rate of going, we expected to reach the land by sundown or soon after; and then our plan was to make a tent of the boat-sail, and sleep under it till morning, when by rising early we thought we could take a run over the island, and perhaps get some fruit and vegetables. By that time, should a light breeze spring up during the night, we thought it likely the ship would be well up by the land, and we could pull out and get on board without difficulty. Before long we found that distances are very deceptive in these latitudes where the atmosphere is so clear; for notwithstanding the statement of the captain that by the reckoning we were seventy miles from land, we believed that he only told us so to deter us from going, and that we were not much more than half that distance. In rowing we made a division of our number, taking turns or watches of an hour each at the oars, so as to share the labor. Once fairly under way, with a smooth sea and a pleasant day before us, we became exceedingly merry at the expense of our fellow-passengers whom we had left in the ship to drift about in the calm, and it afforded us much diversion to think how they would be disappointed upon finding that we were in earnest about going ashore. Before long we had cause to wish ourselves back again in the ship, which goes to prove that apparently the most unfortunate are often less so than those who seem to be favored by circumstances.

At noon we took a lunch, and refreshed ourselves with a drink of water all round. We had also a good supply of cigars, which we smoked with great relish after our pull; and I think there never was a happier set than we were for the time. Still there was but a single peak on the horizon. It was blue and dim in the distance, and apparently not much higher than when we saw it from the mast-head, from which we inferred that there must be a current setting against us. The Anteus was hull down, yet we seemed as far from the land as when we started.

A ripple beginning to show upon the water, we hoisted our sail to catch the breeze, and found that it helped us one or two knots an hour. With songs and anecdotes we passed the time pleasantly till 3 P.M., when we entirely lost sight of the vessel. Paxton, the whaleman, now stood up in the boat to take an observation of the land. There were a few more peaks in sight; the middle peak, which was the first we made, began to loom up very plainly, showing a flat top. It was the mountain called Yonka, which is said to be three thousand feet high. We were apparently forty miles yet from the nearest point; and the sun setting here in May at a little after five, we began to feel uneasy concerning the weather, which showed signs of a change. All of us, having gone so far, were in favor of keeping on, though in secret we thought there was a good deal of danger. At sunset we took another observation. The land had risen quite over the water from end to end, and we hoped to reach it in about three hours. It is true none of us knew any thing about the shores, whether they abounded in bays or not, and if so where any safe place of landing could be found, which made us doubtful how to steer. Clouds were gathering all over the horizon; a few stars shone out dimly overhead, and the shades of night began to cover the island as with a shroud. Swiftly, yet with resistless power, the clouds swept over the whole sky, and the horizon, in all the grandeur of its vast circle, was lost in the shades of night. No sail was near; no light shone upon us now but the dim rays of a few solitary stars through the rugged masses of clouds; no sound broke upon the listening ear save the weary stroke of our oars: a gloom had settled upon the mighty wilderness of waters, and we were awed and silent, for we knew that the spirit of God was there, and darkness was his secret place; that "his pavilion round about him were dark waters and thick clouds of the skies."

One large black mass of clouds rose up on the weather quarter; a low moaning came over the sea, and the air became suddenly chill, and the waters rippled around us, and were tossed about by the unseen Power, and we trembled, for we beheld the coming of the storm that was soon to burst upon us in all the majesty of its wrath. For a while there was the stillness of death; then "the Lord thundered in the heavens, and the Highest gave his voice," and out of the darkness came the storm. In fierce and sudden gusts it came, terrible in its resistless might; lashing the sea into a white foam, tossing and whirling overhead, with its thousand arms outstretched; grasping up the waters as it raged over the deep, and scourging them madly through the air, while it moaned and shrieked like the dread spirit of desolation.

BOAT IN A STORM.

Every one of us cowered down in the boat to keep her balanced. The spray washed over us fearfully, and the sail shook so in the wind, having let go all, that we thought it would tear the mast out. At this time we were about three leagues from the S.E. end of the island, which was the nearest point then in sight. As the cloud spread by the attraction of the land, the whole island became wrapped in a dark shroud of mist, and in half an hour we could discern nothing but the gloom of the storm around us, as we bore down toward the darkest part on the lea. Our lamp was now quenched by a heavy sea, and being unable to distinguish the points of the compass, we were fearful we should miss the island and be carried off so far that we could never reach it again. Whenever there was a lull we tried to haul in our sheet, but a sudden flaw striking us once, the boat lay over till she buried her gunwales, and the sea broke heavily over her lee side, and the crew at the same time springing in a body to the weather side, to balance her, brought her over suddenly, so that it was a miracle we were not capsized, which, had it happened so far out at sea in the darkness, would have made an end of us. Indeed, it was as much as we could do, by baling continually, to keep her afloat, and every moment we expected to be buried in a watery grave. For the reason that we feared the tide or current which set against us might carry us off beyond reach of the land, we kept up our sail as long as we could, thinking that while we made headway toward the lee of the island we increased our chance of safety. Moreover, we knew it was four hundred miles to the coast of Chili, and we had neither water nor provisions left. At best our position was perilous. Ignorant of the bearings of the harbor, we were at a loss what to do even if we should be able to reach the lee of the island, for we had seen that it was chiefly rock-bound and inaccessible to boats.

About 2 A.M., as well as we could judge, we found ourselves close in under the lee of a high cliff, upon the base of which the surf broke with a tremendous roar. Some three or four of the party, reckless of the consequences, were in favor of running straight in, and attempting to gain the shore at all hazards. The more prudent of us protested against the folly of this course, well knowing that we would be capsized in the surf and dashed to pieces on the rocks. Here we found the evils of having too many masters in an adventure of this kind, where every man who had a will of his own seemed disposed to use it. However, by mild persuasion, we adjusted the difficulty, and agreed to continue on under the lee, where we were sheltered in some degree from the gale, till we should hit upon some safe harbor, if such there was upon the island. The boat was our only resource in case of being left ashore, and all admitted the necessity of preserving it as long as possible. If we found no harbor, we could lie off a short distance and wait till daylight. This plan was so reasonable that none could object to it. As soon as we were well in by the shore, where the gale was cut off by the mountains, we had a light eddy of air in our favor, which induced us to keep up our sail. We soon found the danger of this. A strong flaw from a gap in the land struck us suddenly, and would have capsized

STRUCK BY A FLAW.

us had we not let go every thing, and clung to the weather gunwale till it was over, when we quickly pulled down the sail and took to the oars.

We could see nothing on our starboard but the wild seas as they rolled off into the darkness; on our larboard, a black perpendicular wall of rocks loomed up hundreds of feet high, reaching apparently into the clouds. Sometimes a part of the outline came out clear, with its rugged pinnacles against the sky, and now and then a fearful gorge opened up as we coasted along, through which the wind moaned dismally. It was a very wild and awful place in the dead of night, being so covered with darkness that we scarce knew where we steered, or how soon we might be dashed to pieces in the surf. Once in a while we stopped to listen, thinking we heard voices on the shore, but it was only the moaning of the tempest upon the cliffs, and the frightful beating of the surf below. We seemed almost to be able to touch the black and rugged wall of rocks that stood up out of the sea, and the shock of the returning waves so jarred the boat at times that we clung to the thwarts, and believed we were surely within the jaws of death. As the voices died away which we thought came out from the cliffs there was a lull in the storm, and nothing but the wail of the surf could be heard, sounding very sad and lonesome in gloom of night. It was a dreary and perpetual dirge for the ill-fated mariners who were buried upon that inhospitable shore; a death-moan that forever rises out of the deep for the souls that are lost, and the hearts that can never be united with those that love them upon earth again. I thought how well it was writ by the poet—

"Oh, Solitude! where are the charms
That sages have seen in thy face?
Better dwell in the midst of alarms,
Than reign in this horrible place."

SHIPWRECKED SAILOR.

Having pulled about twelve miles along the shore from Goat Island, where we first got under the lee, and seeing no sign of a cove or harbor, we began to despair of getting ashore before daylight. In this extremity, Abraham, a ship-neighbor of mine, succeeded in lighting the lantern again, which he held out in his hand from the bow, hoping thereby to cast a light upon the rocks, that we might grope out our way and reach some place of safety; but it only seemed to make the darkness thicker than it was before. We therefore concluded it was best to pull on till we rounded a point some few miles ahead, where we thought there might be a cove. So we put out the light and got Paxton to go in the bow as a look-out, he being the most keen-sighted, from the habit of looking from the mast-head for whales. On turning the point we were startled by a loud cry of "Light, ho!" Every body turned to see where it appeared. It was close down by the water, about three miles distant, within a spacious cove that opened upon us as we turned the point. Paxton's quick eye had descried it the moment we hove round the rock. Greatly rejoiced by this discovery, we pulled ahead with a good will and rapidly bore down toward the light.

Chilled through with the sharp gusts from the mountains, wet with spray, and very hungry, we congratulated ourselves that there were still inhabitants on the island, and we could not but think they would give us something to eat, and furnish us with some place of shelter. Captain Brooks had told us that he had been here several times in a whaler; that sometimes people lived upon the island from the coast of Chili, and sometimes it was entirely deserted. The Chilians who frequented this lonely island we knew to be a very bad set of people, chiefly convicts and outcasts, who would not hesitate to rob and murder any stranger whom misfortune or the love of adventure might cast in their power. Pirates, also, had frequented its bays from the time of the buccaneers; and it was a question with us whether the light was made by these outlaws, or by some unfortunate shipwrecked sailors or deserters from some English or American whaleship. The better to provide against danger, we loaded our two guns, and placed them in the bow, as also the harpoon; upon which we steered for the light. All of a sudden it disappeared, as if quenched by water. This was a new source of trouble. What could it mean? There was no doubt we had all seen it. The early voyagers had often seen strange lights at night on the tops of the mountains, which they attributed to supernatural causes; but this was close down by the water, and was too well defined and too distinctly visible to us all either to be a supernatural visitation or the result of some volcanic eruption. While we lay upon our oars wondering what it meant, it again appeared, brighter than before. Now, if the inhabitants were not pirates or freebooters, why did they pursue this mysterious conduct? We suspected that they heard our oars, and had lit a fire on the beach to guide us ashore; but if they wanted us to land in the right place, why did they put out the light and start it up again so strangely? For half an hour it continued thus to disappear and reappear at short intervals in the same mysterious way, for which none of us could account.

It being now about four o'clock in the morning, we felt so cast down by fatigue and dread of death, that we decided to run in at all hazards, and, if necessary, make our way through the breakers. All hands fell to upon the oars, and soon the light bore up again close on by the head. Paxton, who was in the bow, quickly started up, and began peering sharply through the gloom. "What's that?" said he: "look there, my lads. I see something black; don't you see it—there, on the larboard—it looks to me like the hull of a ship! Pull, my lads, pull!" and so all gave way with a will, and in a few minutes the tall masts of a vessel loomed up against the sky within a hundred yards! I shall never forget the joy of the whole party at that sight. The light which we had seen came from a lamp that swung in the lower rigging, and though the ship might be a Chilian convict vessel, or some other craft as little likely to give us a pleasant reception, yet we were too glad to think of that, and straightway pulled up under her stern and hailed her. For a moment there was a pause as our voices broke upon the stillness; then there was a stir on deck, and a voice answered us in clear sailor-like English, "Boat ahoy! where are you from?" "The ship Anteus," said we, "bound for California; what ship is this?" "The Brooklyn, of New York, bound for California. Come on board!"

No longer able to suppress our joy, we gave vent to three hearty cheers—cheers so loud and genuine that they swept over the waters of Juan Fernandez, and went rolling up the valleys in a thousand echoes. In less than five minutes we were all on deck, thankful for our providential deliverance from the horrors of that eventful night.

CHAPTER II.

FIRST IMPRESSIONS OF THE ISLAND.

THE decks of the Brooklyn presented a strange and half-savage scene. Most of the passengers, aroused from their sleep by the shouts of the officers and crew, had rushed upon deck nearly naked, and quite at a loss to know what had happened. While we were answering some of their questions, Captain Richardson, the master, pushed his way through the crowd and asked what all the noise was about. We speedily explained how we had left the Anteus seventy miles out at sea, and how, through the aid of Providence, we had made our way into the harbor and descried the ship's lamp; declaring at the same time our belief that, had we missed the ship, in all probability we would have been dashed to pieces upon the rocks. We then made ourselves known personally to the captain, who was well acquainted with some of the party. He cordially welcomed us on board, and invited us into his cabin, where we gave him a more detailed account of our adventure. Meantime the cook was ordered to get us some breakfast as soon as possible, and Captain Richardson offered us dry clothes, and administered to our wants in the kindest manner. Nor was it long till we felt exceedingly comfortable considering the previous circumstances. We soon had breakfast, which, after our toils and troubles, was truly a God-send. Some of the finest fish I ever ate was on the table; excellent ham and potatoes also, fresh bread, and coffee boiling hot. It was devoured with a most uncommon relish, as you may suppose; and it was none the less agreeable for being seasoned with pleasant conversation.

JUAN FERNANDEZ.

The captain admitted that in all his seafaring career he had never known of any thing more absurd than our adventure, and that it was a miracle we were not every one lost. All the passengers crowded around us as if we had risen from the depths of the sea, and I fancied they examined us as if they had an idea that we were some kind of sea-monsters.

The Brooklyn lay at anchor about half a mile from the boat-landing. At the dawn of day I was on the deck, looking eagerly toward the island. I may as well confess at once that no child could have felt more delighted than I did in the anticipation of something illusive and enchanting. My heart throbbed with impatience to see what it was that cast so strange a fascination about that lonely spot. All was wrapped in mist; but the air was filled with fresh odors of land, and wafts of sweetness more delicious than the scent of new-mown hay. The storm had ceased, and the soft-echoed bleating of goats, and the distant baying of wild dogs were all the sounds of life that broke upon the stillness. It seemed as if the sun, loth to disturb the ocean in its rest, or reveal the scene of beauty that lay slumbering upon its bosom, would never rise again, so gently the light stole upon the eastern sky, so softly it absorbed the shadows of night. I watched the golden glow as it spread over the heavens, and beheld at last the sun in all his majesty scatter away the thick vapors that lay around his resting-place, and each vale was opened out in the glowing light of the morning, and the mountains that towered out of the sea were bathed in the glory of his rays.

Never shall I forget the strange delight with which I gazed upon that isle of romance; the unfeigned rapture I felt in the anticipation of exploring that miniature world in the desert of waters, so fraught with the happiest associations of youth; so remote from all the ordinary realities of life; the actual embodiment of the most absorbing, most fascinating of all the dreams of fancy. Many foreign lands I had seen; many islands scattered over the broad ocean, rich and wondrous in their romantic beauty; many glens of Utopian loveliness; mountain heights weird and impressive in their sublimity; but nothing to equal this in variety of outline and undefinable richness of coloring; nothing so dreamlike, so wrapped in illusion, so strange and absorbing in its novelty. Great peaks of reddish rock seemed to pierce the sky wherever I looked; a thousand rugged ridges swept upward toward the centre in a perfect maze of enchantment. It was all wild, fascinating, and unreal. The sides of the mountains were covered with patches of rich grass, natural fields of oats, and groves of myrtle and pimento. Abrupt walls of rock rose from the water to the height of a thousand feet. The surf broke in a white line of foam along the shores of the bay, and its measured swell floated upon the air like the voice of a distant cataract. Fields of verdure covered the ravines; ruined and moss-covered walls were scattered over each eminence; and the straw huts of the inhabitants were almost imbosomed in trees, in the midst of the valley, and jets of smoke arose out of the groves and floated off gently in the calm air of the morning. In all the shore, but one spot, a single opening among the rocks, seemed accessible to man. The rest of the coast within view consisted of fearful cliffs overhanging the water, the ridges from which sloped upward as they receded inland, forming a variety of smaller valleys above, which were strangely diversified with woods and grass, and golden fields of wild oats. Close to the water's edge was the dark moss-covered rock, forever moist with the bright spray of the ocean, and above it, cleft in countless fissures by earthquakes in times past, the red burnt earth; and there were gorges through which silvery springs coursed, and cascades fringed with banks of shrubbery; and still higher the slopes were of a bright yellow, which, lying outspread in the glow of the early sunlight, almost dazzled the eye; and round about through the valleys and on the hill-sides, the groves of myrtle, pimento, and corkwood were draped in green, glittering with rain-drops after the storm, and the whole air was tinged with ambrosial tints, and filled with sweet odors; nothing in all the island and its shores, as the sun rose and cast off the mist, but seemed to

"suffer a sea-change
Into something rich and strange."

CHAPTER III.

GOING ASHORE.

No longer able to control our enthusiam, we sprang into the boat and pushed off for the landing. Captain Richardson, who was well acquainted with the ruins of the Chilian settlement, joined us in our intended excursion, and we were accompanied also by a few sporting passengers from the Brooklyn in another boat. The waters of the bay are of crystal clearness; we saw the bottom as we dashed over the swell, at a depth of several fathoms. It was alive with fish and various kinds of marine animals, of which there are great quantities about these shores. Can you conceive, ye landsmen who dwell in cities, and have never buffeted for weary months the gales of old ocean, the joy of once more touching the genial earth when it has become almost a dreamy fancy in the memories of the past! Then think, without a smile of disdain, what a thrill of delight ran through my blood as I pressed my feet for the first time upon the fresh sod of Juan Fernandez! Think of it, too, as the realization of hopes which I had never ceased to cherish from early boyhood; for this was the abiding place, which I now at last beheld, of a wondrous adventurer whose history had filled my soul years ago with indefinite longings for sea-life, shipwreck, and solitude! Yes, here was verily the land of Robinson Crusoe; here, in one of these secluded glens, stood his rustic castle; here

CRUSOE'S CASTLE.

he fed his goats and held converse with his faithful pets; here he found consolation in the devotion of a new friend, his true and honest man Friday; beneath the shade of these trees he unfolded the mysteries of Divine Providence to the simple savage, and proved to the world that there is no position in life which may not be endured by a patient spirit and an abiding confidence in the goodness and mercy of God.

Pardon the fondness with which I linger upon these recollections, reader, for I was one who had fought for poor Robinson in my boyish days as the greatest hero that ever breathed the breath of life; who had always, even to man's estate, secretly cherished in my heart the belief that Alexander the Great, Julius Cæsar, and all the warriors of antiquity were commonplace persons compared with him; that Napoleon Bonaparte, the Duke of Wellington, Colonel Johnson, Tecumseh, and all the noted statesmen and warriors of modern times, were not to be mentioned in the same day with so extraordinary a man; I, who had always regarded him as the most truthful and the very sublimest of adventurers, was now the entranced beholder of his abiding place—walking, breathing, thinking, and seeing on the very spot! There was no fancy about it—not the least; it was a palpable reality! Talk of gold! Why, I tell you, my dear friends, all the gold of California was not worth the ecstatic bliss of that moment!

CRUSOE AT HOME.

CHAPTER IV.

CONDITION OF THE ISLAND IN 1849.

We first went up to a bluff, about half a mile from the boat-landing, where we spent an hour in exploring the ruins of the fortifications built by the Chilians in 1767. There was nothing left but the foundation and a portion of the ramparts of the principal fort, partly imbedded in banks of clay, and neatly covered with moss and weeds. It was originally strongly built of large stones, which were cast down in every direction by the terrible earthquake of 1835; and now all that remained perfect was the front wall of the main rampart and the groundwork of the fort. Not far from these ruins we found the convict cells, which we explored to some extent.

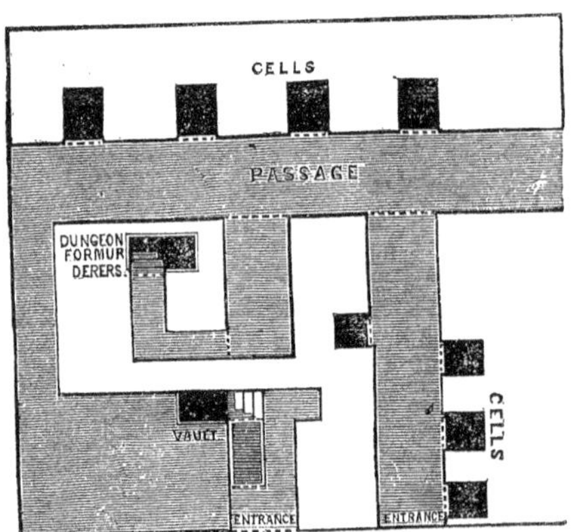

PLAN OF THE CONVICT CELLS.

These cells are dug into the brow of a hill, facing the harbor, and extend underground to the distance of several hundred feet, in the form of passages and vaults, resembling somewhat the Catacombs of Rome. During the penal settlement established here by the Chilian government, the convicts, numbering sometimes many hundreds, were confined in these gloomy dungeons, where they were subjected to the most barbarous treatment. The gates or doors by which the entrances were secured had all been torn down and destroyed, and the excavations were now occupied by wild goats, bats, toads, and different sorts of vermin. Rank fern hung upon the sides; overhead was dripping with a cold and deathlike sweat, and slimy drops coursed down the weeds, and the air was damp and chilly; thick darkness was within in the depths beyond—darkness that no wandering gleam from the light of day ever reached, for heaven never smiled upon those dreary abodes of sin and sorrow. A few of the inner dungeons, for the worst criminals, were dug still deeper underground, and rough stairways of earth led down into them, which were shut out from the upper vaults by strong doors. The size of these lower dungeons was not more than five or six feet in length by four or five in height, from which some idea may be formed of the sufferings endured by the poor wretches confined in them, shut out from the light of heaven, loaded with heavy irons, crushed down by dank and impenetrable walls of earth, starved and beaten by their cruel guards, with no living soul to pity them in their woe, no hope of release save in death. We saw, by the aid of a torch, deep holes scratched in one of the walls, bearing the impression of human fingers. It might have been that some unhappy murderer, goaded to madness by such cruel tortures of body and terrible anguish of mind as drive men to tear even their own flesh when buried before the vital spark is extinct, had grasped out the earth in his desperation, and left the marks in his death agonies upon the clay that entombed him, to tell what no human heart but his had suffered there, no human ear had heard, no human eye had witnessed. The deep, startling echo breaking upon the heavy air, as we sounded the walls, seemed yet to mingle with his curses, and its last sepulchral throb was like the dying moan of the maniac.

Some time before the great earthquake, which destroyed the fortifications and broke up the penal colony, a gang of convicts, amounting to three hundred, succeeded in liberating themselves from their cells. Unable to

CONVICT CELLS.

endure the cruelties inflicted upon them, they broke loose from their chains, and, rushing upon the guards, murdered the greater part of them, and finally seized the garrison. For several days they held complete possession of the island. A whale-ship, belonging to Nantucket, happening to come in at the time for wood and water, they seized the captain, and compelled him to take on board as many of them as the vessel could contain. About two hundred were put on board. They then threatened the captain and officers with instant death in case of any failure to land them on the coast of Peru, whither they determined to go in order to escape the vengeance of the Chilian government. Desirous of getting rid of them as soon as possible, the captain of the whaler ran over for the first land on the coast of Chili, where he put them ashore, leaving them ignorant of their position until they were unable to regain the vessel. They soon discovered that they were only thirty miles from Valparaiso; but, short as the distance was from the Chilian authorities, they evaded all attempts to capture them, and eventually joined the Peruvian army, which was then advancing upon Santiago. The remainder of the prisoners left upon the island escaped in different vessels, and were scattered over various parts of the world. Only a few of the entire number engaged in the massacre were ever captured; sentence of death was passed upon them, and they were shot in the public plaza of Santiago.

Turning our steps toward the settlement of the present residents, we passed a few hours very agreeably in rambling about among their rustic abodes. The total number of inhabitants at this period (1849) is sixteen, consisting of William Pearce, an American, and four or five Chilian men, with their wives and children. No others have lived permanently upon the island for several years. There are in all some six or seven huts, pleasantly surrounded by shrubbery, and well supplied with water

CHILIAN HUTS.

from a spring. These habitations are built of the straw of wild oats, interwoven through wattles or long sticks, and thatched with the same, and, whether from design or accident, are extremely picturesque. The roofs project so as to form an agreeable shade all round; the doorways are covered in by a sort of projecting porch, in the style of the French cottages along the valley of the Seine; small out-houses, erected upon posts, are scattered about each inclosure; and an air of repose and freedom from worldly care pervades the whole place, though the construction of the houses and mode of living are evidently of the most primitive kind. Seen through the green shrubberies that abound in every direction, the bright yellow of the cottages, and the smoke curling up in the still air, have a very cheerful effect; and the prattling voices of the children, mingled with the lively bleating of the kids, and the various pleasant sounds of domestic life, might well lead one to think that the seclusion of these islanders from the busy world is not without its charms. Small patches of ground, fenced with rude stone walls and brushwood, are attached to each of these primitive abodes; and rustic gateways, overrun with wild and luxuriant vines, open in front. Very little attention, however, appears to be bestowed upon the cultivation of the soil; but it looks rich and productive, and might be made to yield abundant crops by a trifling expenditure of labor. The Chilians have never been distinguished for industry; nor is there any evidence here that they depart from their usual philosophy in taking the world easy. Even the American seemed to have caught the prevailing lethargy, and to be content with as little as possible. Vegetables of various kinds grow abundantly wherever the seeds are thrown, among which I noticed excellent radishes, turnips, beets, cabbages, and onions. Potatoes of a very good quality, though not large, are grown in small quantities; and, regarding the natural productiveness of the earth, there seemed to be no reason why they should not be cultivated in sufficient quantities to supply the demands of vessels touching for supplies, and thereby made a profitable source of revenue to the settlers. The grass and wild oats grow in wonderful luxuriance in all the open spaces, and require little attention; and such is the genial character of the climate, that the cattle, of which there seems to be no lack, find ample food to keep them in good condition both in winter and summer. Fig-trees, bearing excellent figs, and vines of various sorts, flourish luxuriantly on the hill-sides. Of fruits there is quite an abundance in the early part of autumn. The peaches were just out of season when we arrived, but we obtained a few which had been peeled and dried in the sun, and we found them large and of excellent flavor. Many of the valleys abound in natural orchards, which have sprung from the seeds planted there by the early voyagers, especially by Lord Anson, who appeared to have taken more interest in the cultivation and settlement of the island than any previous navigator. The disasters experienced by the vessels of this distinguished adventurer in doubling Cape Horn caused him to make Juan Fernandez a rendezvous for the recruiting of his disabled seamen, and for many months he devoted his attention to the production of such vegetables and fruits as he found useful in promoting their recovery; and having likewise in view the misfortunes and necessities of those who might come after him, he caused to be scattered over the island large quantities of seeds, so that, by their increase, abundance and variety of refreshments might be had by all future voyagers. He also left ashore many different sorts of domestic animals, in order that they might propagate and become general throughout the island, for the benefit of shipwrecked mariners, vessels in distress for provisions, and colonists who might hereafter form a settlement there. The philanthropy and moral greatness of these benevolent acts, from which the author could expect to derive little or no advantage during life, can not be too highly commended. If posthumous gratitude can be regarded as a reward, Lord Anson has a just claim to it. How many lives have been saved; how many weather-worn mariners, bowed down with disease, have been renewed in health and strength; how many unhappy castaways have found food abundantly where all they could expect was a lingering death, and have been sustained in their exile, and restored at last to their friends and kindred, through the unselfish benevolence of this brave and kind-hearted navigator, no written record exists to tell; but there are records graven upon the hearts of men that are read by an omniscient eye—a history of good deeds and their reward, more eloquent than human hand hath written.

Besides peaches, quinces, and other fruits common in temperate climates, there is a species of palm called *Chuta*, which produces a fruit of a very rich flavor. Among the different varieties of trees are corkwood, sandal, myrtle, and pimento. The soil in some of the valleys on the north side is wonderfully rich, owing to deposits of burnt earth and decayed vegetable matter washed down from the mountains. There is but little level ground on the island; and although the area of tillable soil is small, yet by the culture of vineyards on the hill-sides, the grazing of sheep and goats on the mountain steeps, and the proper cultivation of the arable valleys, a population of several thousand might subsist comfortably. Pearce, the American, who had thoroughly explored every part of the island, told me he had no doubt three or four thousand people could subsist here without any supply of provisions from other countries. A ready traffic could be established with vessels passing that way, by means of which potatoes, fruits, and other refreshments could be bartered for groceries and clothing. Herds of wild cattle now roam over these beautiful valleys; fine horses may be seen prancing about in gangs, with all the freedom of the mustang; goats in numerous flocks abound among the cliffs; pigeons and other game are abundant; and wild dogs are continually prowling around the settlement.

The few inhabitants at present on the island subsist chiefly upon fish, vegetables, and goat-flesh, of which they have an ample supply. Boat-loads of the finest cod, rockfish, cullet, lobsters, and lamprey eels can be caught in a few hours all around the shores of Cumberland Bay, and doubtless as plentifully in the other bays. Nothing more is necessary than merely the trouble of hauling them out of the water. We fished only for a short time, and nearly filled our boat with the fattest fish I ever saw. Had I not tested myself a fact told me by some of the passengers of the Brooklyn regarding the abundance of the smaller sorts of fish, I could never have believed it—that they will nibble at one's hand if it be put in the water alongside the boat, and a slight ripple made to attract their attention. This is a remarkable truth, which can be attested by any person who has visited these shores and made the experiment. There is no place among the cliffs where goats may not be seen at all times during the day. They live and propagate in the caves, and find sufficient browsing throughout the year in the clefts of the rocks. Lord Anson mentions that some of his hunting parties killed goats which had their ears slit, and they thought it more than probable that these were the very same goats marked by Alexander Selkirk thirty years before; so that it is not unlikely there still exist some of the direct descendants of the herds domesticated by the original Crusoe. The residents of Cumberland Bay have about their huts a considerable number of these animals, tamed, for their milk. When they wish for a supply of goat-flesh or skins (for they often kill them merely for their skins), they go in a body to Goat Island, where they surround the goats and drive them over a cliff into the sea. As soon as they have driven over a sufficient number they take to their boat again, and catch them in the water. Some of them they bring home alive, and keep them till they require fresh meat. Nor are these people destitute of the rarer luxuries of life. By furnishing whale-ships that touch for supplies of water and vegetables with such productions as they can gather up, they obtain in exchange coffee, ship-bread, flour, and clothing; and lately they have been doing a good business in rowing the passengers ashore from the California vessels, and selling them goatskins and various sorts of curiosities. They also charge a small duty for keeping the spring of water clear and the boat-landing free from obstructions, and sometimes obtain a trifle in the way of port charges, in virtue of some pretended authority from the government of Chili.

The shores of Juan Fernandez abound in many different kinds of marine animals, among which the chief are seals and walruses. Formerly sealing vessels made it an object to touch for the purpose of capturing them, but of late years they have become rather scarce, and at present few, if any, vessels visit the island for that purpose.

WALRUS, OR SEA LION.

Situated in the latitude of 33° 40′ S., and longitude 79° W., the climate is temperate and salubrious—never subject to extremes either of heat or cold. In the valleys fronting north, the temperature seldom falls below 50° Fahr. in the coldest season. Open at all times to the pleasant breezes from the ocean, without malaria or any thing to produce disease, beautifully diversified in scenery, and susceptible of being made a convenient stopping-place for vessels bound to the great northwestern continent, it would be difficult to find a more desirable place for a colony of intelligent and industrious people, who would cultivate the land, build good houses, and turn to advantage all the gifts of Providence which have been bestowed upon the island.

The only material drawback is the want of a large and commodious harbor, in which vessels could be hauled up for repairs. This island could never answer any other purpose than that of a casual stopping-place for vessels in want of refreshments, and for this it seems peculiarly adapted. The principal harbors are Port English, on the south side, visited by Lord Anson in 1741; Port Juan, on the west; and Cumberland Bay, on the north side. The latter is the best, and is most generally visited, in consequence of being on the fertile side of the island, where water also is most easily obtained. None of them afford a very secure anchorage, the bottom being deep and rocky; and vessels close to the shore are exposed to sudden and violent flaws from the mountains, and the danger of being driven on the rocks by gales from the ocean. In Cumberland Bay, however, there are places where vessels can ride in safety, by choosing a position suitable to the prevailing winds of the season. The chart and soundings made by Lord Anson will be found useful to navigators who design stopping at Juan Fernandez.

CHAPTER V.

ROBINSON CRUSOE'S CAVE.

Our next expedition was to Robinson Crusoe's Cave. How it obtained that name I am unable to say. The people ashore spoke of it confidently as the place where a seafaring man had lived for many years alone; and I believe most mariners who have visited the island have fixed upon that spot as the actual abode of Alexander Selkirk. There are two ways of getting to the cave from the regular boat-landing; one over a high chain of cliffs, intervening between Crusoe's Valley, or the valley of the cave, and the Chilian huts near the landing; the other by water. The route by land is somewhat difficult; it requires half a day to perform it, and there is danger of being dashed to pieces by the loose earth giving way. In many parts of the island the surface of the cliffs is composed entirely of masses of burnt clay, which upon the slightest touch are apt to roll down, carrying every thing with them. Numerous cases are related by the early voyagers of accidents to seamen and others, in climbing over these treacherous heights. The distance by water is only two miles, and by passing along under the brow of the cliffs a very vivid idea may be had of their strange and romantic formation. We had our guns with us, which we did not fail to use whenever there was an opportunity; but the game, consisting principally of wild goats, kept so far out of reach on the dizzy heights, that they passed through the ordeal in perfect safety. Some of us wanted to go by land and shoot them from above, thinking the bullets would carry farther when fired downward than they seemed to carry when fired from below. The rest of the party had so little confidence in our skill that they dissuaded us from the attempt, on the pretense that the ship might heave in sight while we were absent.

A pleasant row of half an hour brought us to the little cove in Crusoe's Valley. The only landing-place is upon an abrupt bank of rocks, and the surf breaking in at this part of the shore rather heavily, we had to run the boat up in regular beach-comber style. Riding in on the back of a heavy sea, we sprang out as soon as the boat struck, and held our ground, when, by watching our chance for another good sea, we ran her clear out of the water, and made her fast to a big rock for fear she might be carried away. About two hundred yards from where we landed we found the cave.

It lies in a volcanic mass of rock, forming the bluff or termination of a rugged ridge, and looks as if it might be the doorway into the ruins of some grand old castle. The height of the entrance is about fifteen feet, and the distance back into the extremity twenty-five or thirty. It varies in width from ten or twelve to eighteen feet. Within the mouth the surface is of reddish rock, with

CRUSOE'S CAVE.

holes or pockets dug into the sides, which it is probable were used for cupboards by the original occupant. There were likewise large spike-nails driven into the rock, upon which we thought it likely clothing, guns, and household utensils might have been hung even at as remote a date as the time of Selkirk, for they were very rusty, and bore evidence of having been driven into the rock a long time ago. A sort of stone oven, with a sunken place for fire underneath, was partly visible in the back part of the cave, so that by digging away the earth we uncovered it, and made out the purpose for which it was built. There was a darkish line, about a foot wide, reaching up to the roof of the cave, which, by removing the surface a little, we discovered to be produced originally by smoke, cemented in some sort by a drip that still moistened the wall, and this we found came through a hole in the top, which we concluded was the original chimney, now covered over with deposits of earth and leaves from the mountain above. In rooting about the fireplace, so as to get away the loose rubbish that lay over it, one of our party brought to light an earthen vessel, broken a little on one side, but otherwise perfect. It was about eight inches in diameter at the rim, and an inch or two smaller at the bottom, and had some rough marks upon the outside, which we were unable to decipher, on account of the clay which covered it. Afterward we took it out and washed it in a spring near by, when we contrived to decipher one letter and a part of another, with a portion of the date. The rest unfortunately was on the piece which had been broken off, and which we were unable to find, although we searched a long time; for, as may be supposed, we felt curious to know if it was the handiwork of Alexander Selkirk. For my own part, I had but little doubt that this was really one of the earthen pots made by his own hands, and the reason I thought so was that the parts of the letters and date which we deciphered corresponded with his name and the date of his residence, and likewise because it was evident that it must have been imbedded in the ground out of which we dug it long beyond the memory of any living man. I was so convinced of this, and so interested in the discovery, that I made a rough drawing of it on the spot, of which I have since been very glad, inasmuch as it was accidentally dropped out of the boat afterward and lost in the sea.

A RELIC OF CRUSOE.

We searched in vain for other relics of the kind, but all we could find were a few rusty pieces of iron and some old nails. The sides of the cave, as also the top, had marks scattered over them of different kinds, doubtless made there in some idle moment by human hands; but we were unable to make out that any of them had a meaning beyond the unconscious expression of those vague and wandering thoughts which must have passed occasionally through the mind of the solitary mariner who dwelt in this lonely place. They may have been symbolical of the troubled and fluctuating character of his religious feelings before he became a confirmed believer in the wisdom and mercy of Divine Providence, which unhappy state of mind he often refers to in the course of his narrative.

CRUSOE'S DEVOTIONS.

This cave is now occupied only by wild goats and bats, and had not been visited, perhaps, by any human being, until recently, more than once or twice in half a century, and then probably only by some deserter from a whale-ship, who preferred solitude and the risk of starvation to the cruelty of a brutish captain.

In front of the cave, sloping down to the sea-side, is a plain covered with long rank grass, wild oats, radishes, weeds of various kinds, and a few small peach-trees. The latter we supposed were of the stock planted in the island by Lord Anson. From the interior of the cave we looked out over the tangled mass of shrubs, wild flowers, and waving grass in front, and saw that the sea was covered with foam, and the surf beat against the point beyond the cove, and flew up in the air to a prodigious height in white clouds of spray. Large birds wheeled about over the rocky heights, sometimes diving suddenly into the water, from which they rose again flecked with foam, and, soaring upward in the sunlight, their wings seemed to sparkle with jewels out of the ocean. Following the curve of the horizon, the view is suddenly cut off by a huge cliff of lava that rises directly out of the water to the height of twelve or fifteen hundred feet. It forms an abrupt precipice in front, and joins a range of rugged cliffs behind, which all abound in wonderful ledges overlooking the depths below, dark and lonesome caverns, and sharp pinnacles piercing the clonds in every direction. Goat-paths wind around them in places apparently inaccessible, and we saw herds of goats running swiftly along the dizzy heights overhanging the sea, where we almost fancied the birds of the air would fear to fly; they bounded over the frightful fissures in the rocks, and clung to the walls of cliffs with wonderful agility and tenacity of foot, and sometimes they were so high up that they looked hardly bigger than rabbits, and we thought it impossible that they could be goats.

THE VALLEY WITH THE CAVE AND CLIFF.

Looking back into the valley, we beheld mountains stretching up to a hundred different peaks, the sides covered with woods and fields of golden-colored oats, and the ravines fringed with green banks of grass and wild flowers of every hue. A stream of pure spring water rippled down over the rocks, and wound through the centre of the valley, breaking out at intervals into bright cascades, which glimmered freshly in the warm rays of the sun; its margins were fringed with rich grass and fragrant flowers, and groves of myrtle overhung the little lakelets that were made in its course, and seemed to linger there like mirrored beauties spell-bound. Ridges of amber-colored earth, mingled with rugged and moss-covered lava, sloped down from the mountains on every side and converged into the valley, as if attracted by its romantic beauties. Immense masses of rock, cast off from the towering cliffs by some dread convulsion of the elements, had fallen from the heights, and now lay nestling in the very bosom of the valley, enamored with its charms. Even the birds of the air seemed spell-bound within this enchanted circle; their songs were low and soft, and I fancied they hung in the air with a kind of rapture when they rose out of their sylvan homes, and looked down at all the wondrous beauties that lay outspread beneath them.

Some of us scattered off into the woods of myrtle, or lay down by the spring in the pleasant shade of the trees, and bathed our faces and drank of the cool water; others went up the hill-sides in search of peaches, or gathered seeds and specimens of wild flowers to carry home. Too happy in the change, after our gloomy passage round Cape Horn, I rambled up the valley alone, and dreamed glowing day-dreams of Robinson Crusoe. Of all the islands of the sea, this had ever been the paradise of my boyish fancy. Even later in life, when some hard experience before the mast had worn off a good deal of the romance of sea-life, I could never think of Juan Fernandez without a strong desire to be shipwrecked there, and spend the remainder of my days dressed in goatskins, rambling about the cliffs, and hunting wild goats. It

DREAM-LAND CRUSOE.

was a very imprudent desire, to be sure, not at all sensible; but I am now making a confession of facts rather out of the common order, and for which it would be useless to offer any excuse. Pleasant scenes of my early life rose up before me now with all their original freshness. How well I remembered the first time I read the surprising adventures of Robinson Crusoe! It was in the country, where I had never learned the worldly wisdom of the rising generation in cities. Indeed, I had never seen a city, and only knew by hearsay that such wonderful places existed. My father, after an absence of some weeks, returned with an illustrated volume of Crusoe, bound in cream-colored muslin (how plainly I could see that book now!), which he gave me, with a smiling admonition not to commence reading it for two or three years, by which time he hoped I would be old enough to understand it. That very night I was in a new world—a world all strange and fascinating, yet to me as real as the world around me. How I devoured each enchanting page, and sighed to think of ever getting through such a delightful history. It was the first

FAIRY COVE.

book beyond mere fairy tales (which I had almost begun to doubt), the first narrative descriptive of real life that I had ever read. Such a thing as a doubt as to its entire truthfulness never entered my head. I lingered over it with the most intense and credulous interest, and long after parental authority had compelled me to give it up for the night, my whole soul was filled with a confusion of novel and delightful sensations. Before daylight I was up again; I could not read in the dark, but I could open the magic book and smell the leaves fresh from the press; and before the type was visible I could trace out the figures in the prints, and gaze in breathless wonder upon the wild man in the goatskins.

The big tears stood in my eyes when I was through;

but I found consolation in reading it again and again; in picturing out a thousand things that perhaps De Foe never dreamt of; and each night when I went to bed I earnestly prayed to God that I might some day or other be cast upon a desolate island, and live to become as wonderful a man as Robinson Crusoe. Yet, not content with that, I devoted all my leisure hours to making knife-cases, caps, and shot-pouches out of rabbit-skins, in the faint hope that it would hasten the blissful disaster. Years passed away; I lived on the banks of the Ohio; I had been upon the ocean. Still a boy in years, and more so perhaps in feeling, the dream was not ended. I gathered up drift-wood, and built a hut among the rocks; whole days I lay there thinking of that island in the far-off seas. A piece of tarred plank from some steam-boat had a sweeter scent to me than the most odorous flower; for, as I lay smelling it by the hour, it brought up such exquisite visions of shipwreck as never before, perhaps, so charmed the fancy of a dreaming youth. Well I remembered, too, the favored few that I let into the secret; how we went every afternoon to a sand-bar, and called it Crusoe's Island; how I was Robinson Crusoe, and the friend of my heart Friday, whom I caused to be painted from head to foot with black mud, as also the rest of my

RESCUE OF FRIDAY.

friends; and then the battles we had; the devouring of the dead men; the horrible dances, and chasing into the water; and, above all, the rescue of my beloved Friday—how vividly I saw those scenes again!

Years passed on; I was a sailor before the mast. Alas! what a sad reality! I saw men flogged like beasts; I saw cruelty, hardship, disease, death in their worst forms; so much I saw that I was glad to take the place of a wandering outcast upon the shores of a sickly island ten thousand miles from home, to escape the horrors of that life. Yet the dream was not ended. Bright and beautiful as ever seemed to me that little world upon the seas, where dwelt in solitude the shipwrecked mariner. In the vicissitudes of fortune, I was again a wanderer; impelled by that vision of island-life which for seventeen years had never ceased to haunt me, I cast all upon the hazard of a die—escaped in an open boat through the perils of a storm, and now—where was I? What pleasant sadness was it that weighed upon my heart? Was all this a dream of youth; was it here to end, never more to give one gleam of joy; was the happy credulity, the freshness, the enthusiasm of boyhood gone forever? Could it be that this was not Crusoe's Valley at last—this spot, which I had often seen in fancy from the banks of the Ohio, dim in the mist of seas that lay between? Did I really wander through it, or was it still a dream?

And where was the king of the island; the hero of my boyish fancy; he who had delighted me with the narrative of his romantic career, as man had never done before, as all the pleasures of life have never done since; where was the genial, the earnest, the adventurous Robinson Crusoe? Could it be that there was no "mortal mixture of earth's mould in him;" that he was barely the simple mariner Alexander Selkirk? No! no! Robinson Crusoe himself had wandered through these very groves of myrtle; he had quenched his thirst in the spring that bubbled through the moss at my feet; had slept during the glare of noon in the shade of those overhanging grottoes; had dreamed his day-dreams in these secluded glens.

CRUSOE ASLEEP.

Here, too, Friday had followed his master; the simple, childlike Friday, the most devoted of servants, the gentlest of savages, the faithfullest of men! Blessing on thee, Robinson, how I have admired thy prolific genius; how I have loved thee for thine honest truthfulness! And blessings on thee, Friday, how my young heart hath warmed toward thee! how I have laughed at thy scalded fingers, and wept lest the savages should take thee away from me! * * *

CHAPTER VI.

THE VALLEY ON FIRE.

THERE was a sudden rustling in the bushes.

"Hallo, there!" shouted a voice. I looked round and beheld a fellow-passenger, a strange, eccentric man, who was seldom known to laugh, and whose chief pleasure consisted in reducing every thing to the practical standard of common sense. He was deeper than would appear at first sight, and not a bad sort of person at heart, but a little wayward and desponding in his views of life.

"You'll catch cold," said he; "nothing gives a cold so quick as sitting on the damp ground."

"True," said I, smiling; "but recollect the romance of the thing."

"Romance," rejoined the sad man, "won't cure a cold. I never knew it to cure one in my life."

"Well, I suppose you're right. Every body is right who believes in nothing but reality. The hewer of wood and the drawer of water gets more credit in the world for good sense than the unhappy genius who affords pleasure to thousands."

"So he ought—he's a much more useful man."

"Granted; we won't dispute so well-established a truism. Now let us cut a few walking-sticks to carry home. It will please our friends to find that we thought of them in this outlandish part of the world."

"To be sure; if you like. But you'll never carry them home. No, sir, you can't do it. You'll lose them before you get half way to America."

"No matter—they cost nothing. Lend me your knife, and we'll try the experiment, at all events."

I then cut a number of walking-sticks and tied them up in a bundle. And here, while the warning of the doubter is fresh in my mind, let me mention the fate of these much-valued relics. I cut four beautiful sticks of myrtle, every one of which I lost before I reached California, though I was very careful where I kept them—so careful, indeed, that I hid them away on board the ship and never could find them again.

On our way back to the cave, as we emerged from the grove, I was astonished to see the entire valley in a blaze of fire. It raged and crackled up the sides of the mountains, blazing wildly and filling the whole sky with smoke. The beautiful valley upon which I had gazed with such delight a few hours before, seemed destined to be laid waste by some fierce and unconquerable destroyer, that devoured trees, shrubs, and flowers in its desolating career. The roar of the mad rushing flames, the seething tongues of fire shooting out from the bowers of shrubbery, the whirling smoke sweeping upward around the pinnacles of rock, the angry sea dimly seen through the chaos, and the sharp screaming of the sea-birds and dismal howling of the wild dogs, impressed me with a terrible picture of desolation. It seemed as if some dreadful convulsion of nature had burst forth soon to cover the island with seething lava or ingulf it in the ocean.

"What can it be?" said I. "Isn't it a grand sight? Perhaps a volcano has broken out. Surely it must be some awful visitation of Providence. It wouldn't be comfortable, however, to be broiled in lava, so I think the sooner we get down to the boats the better."

"There's no hurry," said my friend; "it's nothing but the Californians down at the cave. I told them before I left that they'd set fire to the grass if they kept piling the brush up in that way. Now you see they've done it."

"Yes, I see they have; and a tolerably big fire they've made of it too."

I almost forgave them the wanton act of Vandalism, so sublime was the scene. It was worth a voyage round Cape Horn to see it.

"Plenty of it," muttered the sad man, "to cook all the food that can be raised in these diggings. I wouldn't give an acre of ground in Illinois for the whole island. I only wish they'd burn it up while they're at it—if it be an island at all, which I ain't quite sure of yet."

THE CALIFORNIANS IN JUAN FERNANDEZ.

We reached the cave by rushing through the flames. When we arrived near the mouth, I was amused to find about twenty long-bearded Californians, dressed in red shirts, with leather belts round their bodies, garnished with knives and pistols, and picks in their hands, with which they were digging into the walls of Selkirk's castle in search of curiosities. Their guns were stacked up outside, and several of the party were engaged in cooking fish and boiling coffee. They had battered away at the sides, top, and bottom of the cave in their eager search for relics, till they had left scarcely a dozen square feet of the original surface. Every man had literally his pocket full of rocks. It was a curious sight, here in this solitary island, scarcely known to mariners save as the resort of pirates, deserters, and buccaneers, and chiefly to the reading world at home as the land of Robinson Crusoe, to see these adventurous Americans in their red shirts, lounging about the veritable castle of the "wild

THE CALIFORNIANS IN JUAN FERNANDEZ.

man in the goatskins," digging out the walls, smoking cigars, whittling sticks, and talking in plain English about California and the election of General Taylor. Some of them even went so far as to propose a "prospecting" expedition through Crusoe's Valley in search of gold, while others got up a warm debate on the subject of annexation—the annexation of Juan Fernandez. One long, lank, slab-sided fellow, with a leathern sort of face, and two copious streams of tobacco-juice running down from the corners of his mouth, was leaning on his pick outside the cave, spreading forth his sentiments for the benefit of the group of gentlemen who were cooking the fish.

"I tell you, feller-citizens," said he, aroused into something like prophetic enthusiasm as the subject warmed upon his mind, "I tell you it's manifest destiny. Joo-an

Fernandays is bound by all the rights of con-san-guity to be a part of the great Ree-public of Free States. Gentlemen, I'm a destiny-man myself; I go the whole figure, sir; yes, sir, I'm none of your old Hunkers. I go for Joo-an Fernandays and California, and any other small patches of airth that may be laying around the vicinity. We want 'em all, gentlemen; we want 'em for our whale-ships and the yeomanry of our country! (cheers.) We'll buy 'em from the Spaniards, sir, with our gold; if we can't buy 'em sir, by hokey! we'll TAKE 'EM, sir! (Renewed cheers.) I ask you, gentlemen—I appeal to your feelins as feller-citizens of *thee* greatest concatenation of states on *thee* face of God's airth, are you the men that'll refuse to fight for your country? (Cheers, and cries of No, no, we ain't the men; hurra for Joo-an Fernandays!) Then, by Jupiter, sir, we'll have it! We'll have it as sure as the Star of Empire shines like the bright Loo-min-ary of Destiny in the broad Panoply of Heaven (and more especially in the western section of it). We'll have it, sir, as sure as that redolent and inspiring Loominary beckons us on, sir, like a dazzling joo-el on the pre-monitary finger of Hope; and the glorious Stars and Stripes, feller-citizens, shall wave proudly in the zephyrs of futurity over the exalted peaks of Joo-an Fernandays!" (Tremendous sensation, during which the orator takes a fresh chew of tobacco, and sits down.)

As soon as the party of annexationists perceived us, they called out to us to heave to, and make ourselves at home. "Come on, gentlemen, come on! No ceremony. We're all Americans! this is a free country. Here's fish! here's bread! here's coffee! Help yourselves, gentlemen! This is a great country, gentlemen—a great country!" Of course we fell to work upon the fish, which was a splendid cod, and the bread and the coffee too, and very palatable we found them all, and exceedingly jolly and entertaining the "gentlemen from the Brooklyn." These lively individuals had made the most of their time in the way of enjoying themselves ashore.

About a week before our arrival they gave a grand party in honor of the American nation in general. It was in rather a novel sort of place, to be sure, but none the worse for that—one of the large caves near the boat-landing. On this eventful occasion they "scared up," as they alleged, sundry delicacies from home, such as preserved meats, pound-cake, Champagne, and wines of various sorts, and out of their number they produced a full band of music. They also, by clearing the earth and beating it down, made a very good place for dancing, and they had waltzes, polkas, and cotillons, in perfect ballroom style. It was rather a novel entertainment, take it altogether, in the solitudes of Juan Fernandez. I have forgotten whether the four Chilian ladies of the island attended; if they did not, it was certainly not for want of an invitation. The American Crusoe was there, no longer monarch of all he surveyed. Poor fellow, his reign was over. The Californians were the sovereigns now.

After our snack with the Brooklynites, we joined our comrades down on the beach. They had shot at a great many wild goats, without hitting any, of course. The rest of the afternoon we spent in catching fish for supper.

FISHING.

CHAPTER VII.

THE CAVE OF THE BUCCANEERS.

It now began to grow late, and we thought it best to look about us for some place where we could sleep. Captain Richardson very kindly offered us the use of his cabin, but he was crowded with passengers, and we preferred staying ashore. There was something novel in sleeping ashore, but neither novelty nor comfort in a vessel with a hundred and eighty Californians on board. Brigham and a few others took our boat, and went over near the old fort to search out a camping-ground, while the rest of the party and myself started off with the captain to explore a grotto. We had a couple of sailors to row us, which helped to make the trip rather pleasant.

Turning a point of rocks, we steered directly into the mouth of the grotto, and ran in some forty or fifty feet, till nearly lost in darkness. It was a very wild and rugged place—a fit abode for the buccaneers.

The cliff into which the cave runs is composed of great rocks, covered on top with a soil of red, burned earth. The swell of the sea broke upon the base with a loud roar, and the surf, rolling inward into the depths of the grotto, made a deep reverberation, like the dashing of water under a bridge. There was some difficulty in effecting a landing among these subterranean rocks, which were round and slippery. The water was very deep, and abounded in seaweed. On gaining a dry place, we found the interior quite lofty and spacious, and tending upward into the very bowels of the mountain. Some said there was a way out clear up in the middle of the island. Overhead it was hung with stalactites, some of which were of great size and wonderful formation. Abraham and myself climbed up in the dark about a hundred feet, where we entirely lost sight of the mouth, and could hardly see an inch before us. As we turned back and began to descend, our friends down below looked like gigantic monsters standing in the rays of light near the entrance. I broke off some pieces of rock and put them in my pocket, as tokens of my visit to this strange place.

On reaching the boat again, we found a group of our comrades seated around a natural basin in the rocks, regaling themselves on bread and water. The water, I think, was the clearest and best I ever tasted. It trickled down from the top of the cave, and fell into the basin with a most refreshing sound. I drank a pint gobletful, and found it uncommonly cool and pure. Nothing more remaining to be seen, we started off for the boat-landing, near the huts, where we parted with our friend the captain, and then, it being somewhat late, we went in search of our party.

CHAPTER VIII.

LODGINGS UNDER GROUND.

When we arrived on the ground selected by Brigham and the others, we found that they had made but little progress in cutting wood for the posts, and much remained to be done before we could get up the tent.

Heavy clouds hung over the tops of the mountains; the surf moaned dismally upon the rocks; big drops of rain began to strike us through the gusts of wind that swept down over the cliffs, and there was every prospect of a wet and stormy night. It was now quite dark. After some talk, we thought it best to abandon our plan of sleeping under the sail. Finally, we agreed to go in search of a cave under the brow of a neighboring cliff. We had seen it during the day, and although a very unpromising place, we thought it would serve to protect us against the rain. We therefore took our oars and sail upon our shoulders, together with what few weapons of defense we had, and stumbled about in the dark for some time, till we had the good fortune to find the mouth of the cave. In the course of a few minutes we struck a light by a lucky chance, and then looked in. There seemed to be no bottom to it, and, so far as we could perceive, neither sides nor top. Certainly there was not a living soul about the premises to deny us admission; so we crept down, as we thought, into the bowels of the earth, and, seeing nobody there, took possession of our lodgings, such as they were.

It was a damp and gloomy place enough, reeking with mould, and smelling very strong of strange animals. The rocks hung gaping over our heads, as if ready to fall down upon us at the mere sound of our voices; the ground was covered with dirty straw, left there probably by some deserters from a whale-ship, and all around the sides were full of holes, which we supposed from the smell must be inhabited by foxes, rats, and perhaps snakes, though we were afterward told there were no reptiles on the island. We soon found that there were plenty of spiders and fleas in the straw. The ground being damp, we spread our sail over it, in order to make a sort of bed; and, being in a measure protected by a clump of bushes placed in the entrance by the previous occupants to keep out the wind and rain, we did not altogether despair of passing a tolerably comfortable night.

For a while there was not much said by any body; we were all busy looking about us. Some were looking at the rocks overhead; some into the holes, where they thought there might be wild animals; and myself and a few others were trying to light a fire in the back part of the cave. It smoked so that we had to give it up at last, for it well-nigh stifled the whole party.

By this time, being all tired, we lay down, and had some talk about Robinson Crusoe.

"If he lived in such holes as this," said one, "I don't think he had much sleep."

"No," muttered another, "that sort of thing reads a good deal better than it feels; but there's no telling how a man may get used to it. Eels get used to being skinned, and I've heard of a horse that lived on five straws a day."

"For my part," adds a third, "I like it: there's romance about it—and convenience too, in some respects. For the matter of clothing, a man could wear goatskins. Tailors never dunned Robinson Crusoe. It goes a great way toward making a man happy to be independent of fashion. Being dunned makes a man miserable."

"Yes, it makes him travel a long way sometimes," sighs another, thoughtfully. "I'd be willing to live here a few years to get rid of society. What a glorious thing it must be to have nothing to do but hunt wild goats! Robinson had a jolly time of it; no accounts to make out, no office-hours to keep, nobody to call him to account every morning for being ten minutes too late, in consequence of a frolic. Talking about frolics, he wasn't tempted with liquor, or bad company either; he chose his own company: he had his parrot, his goats, his man Friday—all steady sort of fellows, with no nonsense about them. I'll venture to say they never drank any thing stronger than water."

CRUSOE AND HIS COMRADES.

"No," adds another, gloomily, "it isn't likely they applied 'hot and rebellious liquors to their blood.' But a man who lives alone has no occasion to drink. He has no love affairs on hand to drive him to it."

"Nor a scolding wife. I've known men to go all the way to California to get rid of a woman's tongue."

There was a pause here, as most of the talkers began to drop off to sleep.

"Gentlemen," said somebody in the party, who had been listening attentively to the conversation, "I don't believe a single word of it. I don't believe there ever was such a man as Robinson Crusoe in the world. I don't believe there ever was such a man as Friday. In my opinion, the whole thing is a lie, from beginning to end. I consider Robinson Crusoe a humbug!"

"Who says it's all a lie?" cried several voices, fiercely; "who calls Robinson Crusoe a humbug?"

"That is to say," replied the culprit, modifying the remark, "I don't think the history is altogether true. Such a person might have lived here, but he added something on when he told his story. He knew very well his man Friday, or his dogs and parrots were not going to expose his falsehoods."

"Pooh! you don't believe in any thing; you never did believe in any thing since you were born. Perhaps you don't believe in that. Are you quite sure you are here yourself?"

"Well, to be candid, when I look about me and see what a queer sort of a place it is, I don't feel quite sure; there's room for doubt."

"Doubt, sir! doubt? Do you doubt Friday? Do you think there's room for doubt in him?"

"Possibly there may have been such a man. I say there *may* have been; I wouldn't swear to it."

"Fudge, sir! fudge! The fact is, you make yourself ridiculous. You are troubled with dyspepsia."

"I am rayther dyspeptic, gentlemen, rayther so. I hope you'll excuse me, but I can't exactly say I believe in Crusoe. It ain't my fault—the belief ain't naturally in me."

Upon which, having made this acknowledgment, we let him alone, and he turned over and went to sleep. We now pricked up our lamp, and prepared to follow his example, when a question arose as to the propriety of standing watches during the night—a precaution thought necessary by some in consequence of the treacherous character of the Spaniards. There were eleven of us, which would allow one hour to each person. For my part, I thought there was not much danger, and proposed letting every man who felt uneasy stand watches for himself. We had labored without rest for thirty-six hours, and I was willing to trust to Providence for safety, and make the most of our time for sleeping. A majority being of the same opinion, the plan of standing watches was abandoned; and having loaded our two guns, we placed them in a convenient position commanding the mouth of the cave. I got the harpoon and stood it up near me, for I had made up my mind to fasten on to the first Spaniard that came within reach.

ATTACK OF THE ROBBERS.

Scarcely had we closed our eyes and fallen into a restless doze, when a nervous gentleman in the party rose up on his hands and knees, and cautiously uttered these words:

"Friends, don't you think we'd better put out the light. The Spaniards may be armed, and if they come here, the lamp will show them where we are, and they'll be sure to take aim at our heads."

"Sure enough," whispered two or three at once, "we didn't think of that; they can't see us in the dark, however, unless they have eyes like cats. Let us put out the light, by all means."

So with that we were about to put out the light, when the man who had doubts in regard to Robinson Crusoe rose up on his hands and knees likewise, and said,

"Hold on! I think you'd better not do that. It ain't policy. I don't believe in it myself."

"Confound it, sir," cried half a dozen voices, angrily, "you don't believe in any thing. What's the reason you don't believe in it, eh? What's the reason, sir?"

"Well, I'll tell you why. Because, if you put out the light, we can't see where to shoot. Likely as not we'd shoot one another. If I feel certain of any thing, it is, that I'd be the first man shot; it's my luck. I know I'd be a dead man before morning."

There was something in this suggestion not to be laughed at. The most indignant of us felt the full force of it. To shoot our enemies in self-defense seemed reasonable enough, but to shoot any of our own party, even the man who doubted Robinson Crusoe, would be a very serious calamity. At last, after a good deal of talk, we compromised the matter by putting the lamp under an old hat with a hole in the top. This done, we tried to go to sleep.

Brigham went to the mouth of the cave about midnight to take an observation. He was armed with one of the guns.

"What's that?" said he, sharply; "I hear something! Gentlemen, I hear something! Hallo! who goes there?"

There was no answer. Nothing could be heard but the moaning of the surf down on the beach.

"A Spaniard! by heavens, a Spaniard! I'll shoot him—I'll shoot him through the head!"

"Don't fire, Brigham," said I, for I wanted a chance to fasten on with the harpoon; "wait till he comes up, and ask him what he wants."

"Ahoy there! What do you want? Answer quick, or I'll shoot you! Speak, or you're a dead man!"

All hands were now in commotion. We rushed to the mouth of the cave in a body, determined to defend ourselves to the last extremity.

"Gentlemen," cried Brigham, a little confused, "it's a goat! I see him now, in the rays of the moon; a live goat, coming down the cliff. Shall I kill him for breakfast?"

"Wait," said I, "till he comes a little closer; I'll bend on to him with the harpoon."

"You'd better let him alone," said the Doubter, in a sepulchral voice. "Likely as not it's a tame goat or a chicken belonging to the American down there."

"A tame devil, sir! How do you suppose they could keep tame goats in such a place as this. Your remark concerning the chicken is beneath contempt!"

"Well, I don't know why. Tain't my nature to take an entire goat without proof. I thought it might be a chicken."

"Then you'd better go and satisfy yourself, if you're not afraid."

The Doubter did so. He walked a few steps toward the object, so as to get sight of its outline, and then returned, saying,

"That thing there isn't a goat at all—neyther is it a chicken."

"What is it, then?"

"Nothing but a bush."

"What makes it move?"

"The wind, I suppose. I don't know what else could make it move, for it ain't got the first principle of animal life in it. Bushes don't walk about of nights any more than they do in the daytime. I never did believe in it from the beginning, and I told you so, but you wouldn't listen to me."

We said nothing in reply to this, but returned into the cave and lay down again upon the sail.

CHAPTER IX.

COOKING FISH.

Most of the party were snoring in about ten minutes. For myself, I found it impossible to sleep soundly. The gloomy walls of rock, the strange and romantic situation into which chance had thrown me, the remembrance of what I had read of this island in early youth, the dismal moaning of the surf down on the beach, all contributed to confuse my mind. An hour or two before daylight, I was completely chilled through by the dampness of the ground, and entirely beyond sleep.

I heard some voices outside, and got up to see who was talking. Lest it might be the Spaniards, I took the harpoon with me. At the mouth of one of the convict-cells near by I found four of my comrades, who, unable to pass the time any other way, had lit a fire and were baking some fish. They had dug a hole in the ground, which they lined with flat stones, so as to form a kind of oven; this they heated with coals. Then they wrap-

COOKING IN JUAN FERNANDEZ.

ped up a large fish in some leaves, and put it in; and by covering the top over with fire, the fish was very nicely baked. I think I never tasted any thing more delicate or better flavored. We had an abundant meal, which we relished exceedingly. The smoke troubled us a good deal; but, by telling stories of shipwreck, and wondering what our friends at home would think if they could see us here cooking fish, we contrived to pass an hour or so very pleasantly. I then went back into the cave, and turned in once more upon the sail.

Of course, after eating fish at so unusual an hour, I had a confusion of bad dreams. Perhaps they were visions. In this age of spiritual visitations, it is not altogether unlikely the spirits of the island got possession of me. At all events, I saw Robinson Crusoe dressed in goatskins, and felt him breathe, as plainly as I see this paper and feel this pen. How could I help it? for I actually thought it was myself that had been shipwrecked; that I was the very original Crusoe, and no other but the original; and I fancied that Abraham had turned black, and was running about with a rag tied round his waist, and I called him my man Friday, and fully believed him to be Friday. Sometimes I opened my eyes and looked round the dismal cavern, and clenched my fists, and hummed an old air of former times to try if Robinson had become totally savage in his nature; but it was all the same, there was no getting rid of the illusion.

The dawn of day came. No ship was in sight. The sea was white with foam, and gulls were soaring about over the rock-bound shores. I walked down to a spring and bathed my head, which was hot and feverish for want of rest.

Bright and early we started off on a goat-hunt among the mountains. Several passengers from the Brooklyn, well provided with guns, joined the party, and the enthusiasm was general. It had been my greatest desire, from the first sight of the island, to ascend a high peak between the harbor and Crusoe's Valley, and by following the ridge from that point, to explore as far as practicable the interior. For this purpose, I selected as a companion my friend Abraham, in whose enthusiastic spirit and powers of endurance I had great confidence. He was heartily pleased to join me; so, buckling up our belts, we branched off from the party, who by this time were peppering away at the wild goats. We were soon well up on the mountain. Another adventurer joined us before we reached the first elevation; but he was so exhausted by the effort, and so unfavorably impressed by the frightful appearance of the precipices all round, that he was forced to abandon the expedition and return into the valley. We speedily lost sight of him, as he crept down among the declivities.

THE CLIFF.

The side of the mountain which we were ascending was steep and smooth, and was covered with a growth of long grass and wild oats, which made it very hard to keep the goat-paths; and all about us, except where these snake-like traces lay, was as smooth and sloping as the roof of a house. There was one part of the mountain that sloped down in an almost perpendicular line to the verge of the cliff overhanging the sea, where the abrupt fall was more than a thousand feet, lined with sharp crags. This fearful precipice rose like a wall of solid rock out of the sea, and there was a continual roar of surf at its base. There was no way of getting up any higher without scaling the slope above, which, as I said before, was covered with long grass and oats, that lay upon it like the thatch of a house; and the rain which had fallen during the previous night now made it very smooth. I looked at it, I must confess, with something like dismay, thinking how we were to climb over such a steep place without slipping down over the cliff; when I beheld Abraham, of whom I had lost sight for a time, toiling upward upon it like a huge bear. His outline against the sky reminded me especially of a bear of the grizzly species. I saw that he clung to the roots of the grass with his hands, and dug his toes into the soft earth to keep from sliding back, in case his hold should give way. Committing myself to Providence, I started after him by a shorter cut, grasping hold of the grass by the roots as I went. Every few perches, I stopped to search for a strong bunch of grass, for there was nothing else to hold on by. Some of it was so loose that it gave way as soon as I laid hold of it, and I came near going for want of something to balance me. Six inches of a slide would have sent me twirling over the cliff into the raging surf a thousand feet below. Once, impressed with the terrible idea that I was slipping, I stopped short, and my heart beat till it shook me all over. It was only by lying flat down and seizing the roots of the grass with both hands, while I dug my toes into the sod, that I retained my presence of mind. Indeed, at this place, having turned to look back, I was so struck with horror at the frail tenure upon which my life depended, that I turned partly blind, and a rushing noise whirled through my brain at the thought that I should be no longer able to retain my grasp. If for one moment I lost my consciousness and let go my hold of the grass, I would surely be lost; there was no hope; I must be dashed over the precipice, and go spinning through a thousand feet of space till I struck the rocks below, or was buried in the surf. I lay panting for breath, while every muscle quivered as if it would shake loose my grasp. In the space of five minutes I thought more of death than I had ever thought before. Was this to be my end after all? What would they say on board the ship when I was dead? What would be the distress of my friends and kindred at home when they heard how my mangled body was picked up in the surf, and buried upon this lonely rock-bound island? A thousand thoughts flashed through my brain in succession. Even the happy days of my youth rose up before me now, but the vision was sadly mingled with errors and follies that could never be retrieved. Believing my time had come, I looked upward in my agony, and beheld Abraham, scarcely twenty yards in advance, lying down in the same position, with hands stretched out and dug into the roots of the grass.

"Abraham," said I, "this is terrible!"

"Yes," said he, "a foretaste of death, if nothing worse."

"But how in the world are we to get out of it?"

"I don't know—there seems to be no hope; we can't go back again, that's an absolute certainty. In my opinion, we'll have to stay here till somebody comes for us, which doesn't seem a likely chance just now."

A good rest, however, having inspired us with fresh courage, we resolved upon pushing on. There was a narrow ledge about a hundred yards above us; if we could reach that, we would be safe for the present. By great exertion we got a little above the place where we had lain down; and, the sod beginning to give way as before, we threw ourselves on our faces again, and rested a while. In this way, hanging, as it were, between life and death, we at last reached the ledge. Here we flung ourselves on the solid rock, quite exhausted. Abraham was a brave man, but he now lay gasping for breath, as pale as a ghost. I suppose I looked about the same, for, to tell the honest truth, I was well-nigh scared out of my senses. Certainly all the gold of Ophir could not have induced me to go through the same ordeal again.

There was still above us, about five hundred feet higher, a point or pyramid of volcanic rock, that stood out over the sea in a slanting direction. It was the highest peak in the neighborhood of the coast, and was called the Nipple. We had done nothing yet compared with the ascent of that peak. Both of us looked toward it, and smiled.

"Shall we try it?" said Abraham.

"No," said I, "we never could get up there; it would be perfect folly to try."

"I think not, Luff; it isn't so smooth as the place we have just climbed over. Don't you see there are rocks to hold on to?"

"Yes, but they look as if they'd give way. However, if you say so, we'll make the attempt."

With this, we each drew a long breath, and commenced climbing up the rocks. Sometimes we dug our fingers into the crevices and lifted ourselves up, and sometimes we wound around ledges less than a foot wide, overhanging deep chasms, and were forced to cling to the rough points that jutted out in order to keep our balance. Flocks of pigeons flew startled from their nests, and whirled past us, as if affrighted at the intrusion of man. Herds of wild goats dashed by us also, and ran bleating down into the rugged defiles, where they looked like so many insects. The wind whistled mournfully against the sharp crags, and swept against us in such fierce and sudden gusts that we were sometimes obliged to stop and cling to the rocks with all our might to keep from being blown off. At last we reached the base of the Nipple. This was the wildest place of all. Above us stood the dizzy peak, like the turret of a ruined castle, overlooking the surf at a height of nearly two thousand feet. We now lay down again, breathing hard, and a good deal exhausted. When partly recovered, I looked over the edge toward Crusoe's Valley. It was the grandest sight I ever beheld; rugged cliffs and winding ridges hundreds of feet below; a green valley embowered in shrubbery nestling beneath the heights, all calm and smiling in the warm sunshine; slopes of woodland stretching up in the ravines; a line of white spray from the surf all along the shores, and the boundless ocean outspread in one vast sweep beyond.

"I'll tell you what it is, Luff," said Abraham, "this may be all very fine, but I don't want to try it again."

"Nor I either, Abraham. Isn't it awful climbing?"

"Yes, awful enough; but we must get on the top of that old castle there."

"To be sure," said I, rather doubtfully. "Of course, Abraham; we ought to climb that as a sort of climax. It will make an excellent climax either to ourselves or the adventure."

Saying this, I walked a few steps from the place where we were lying down, to see if there was any way of scaling the Nipple. It appeared to be a huge pile of loose rocks ready to fall to pieces upon being touched. It was about a hundred feet high, and nearly perpendicular all round. There was no part that seemed to me at all accessible. Even the first part or foundation could not be reached without passing over a sharp ridge, steep at both sides, and entirely destitute of vegetation. I was not quite mad enough to undertake such a thing as this without the least hope of success.

"No, Abraham," said I, "we can't do it. I see no way of getting up there."

"Let me take a look," said Abraham, who was always

fertile in discoveries. "I think I see a place that we can climb over, so as to get on that horseback sort of a ridge, and the rest of the way may be easier than we suppose."

He then walked a few steps round a ledge of crumbling rock, and I soon saw him climbing up where it seemed as if there was no possible way of holding on. I actually began to think there was something supernatural in his hands and feet; yet I felt an indescribable dread that he would fall at last. For a while I was in perfect agony; each moment I expected to see him roll headlong over the cliff. Presently I lost sight of him altogether. I thought he had lost his balance, and was dashed to atoms below! Seized with horror, I sat down and groaned aloud. Again I rose and ran to the edge of the cliff, shouting wildly in the faint hope that he was not yet lost. There was no answer but the wail of the winds and the moaning of the surf. While I looked from the depths to the fearful height above, I saw his head rise

ABRAHAM ON THE PEAK.

slowly and cautiously over the top of the Nipple; then his body, and then, with a wild shout of triumph, he stood waving his hat on the summit!

There he stood, a man of stalwart frame, now no bigger than a dwarf against the sky!

I saw him point toward the horizon, and, looking in the direction of his finger, perceived the Anteus about twenty miles off under short sail.

He remained but a few minutes in this perilous position, as I supposed on account of the wind, which was now very strong.

On his return, being unable to get down on the same side, he was forced to creep backward over the ridge, and lower himself by fixing his hands in the crevices to the ledge over the sea, from which he made his way round to the starting-point. When he reached the spot where I stood, he sat down, breathing hard, and looking very pale.

"Luff," said he, "don't go up there. It shook under me like a tree. Every flaw of wind made it sway as if it would topple over."

"Why," said I, "after scaring me out of my wits, it isn't exactly fair to deprive me of some satisfaction."

"Don't do it, Luff; I warn you as a friend! It ought to be satisfaction enough to find me here safe and sound, after such a climb as that."

"No, Abraham, I must do it; because when we return to the ship, don't you see what an advantage you'll have over me?"

"Only in being the greater fool."

"Then there must be two fools, to make us even. It would hardly be friendly to let you be the only one; so here goes, Abraham. In case I tumble over, give my love to all at home, and tell them I died like a Trojan."

All this was folly, to be sure; but how could I help it? how could I bear the thought of hearing Abraham talk about having scaled the Nipple, while I was ingloriouly groaning for him down below? It would mortify me to the very soul.

Following now the same path that taken, I was soon on top of the first el. ing lighter and more active, though not so strong, I had rather the advantage in climbing. Here I wound round by a different way, so as to reach the ridge that led over the chasm. It was about the width of a horse's back, sloping down abruptly on each side. The distance was not over twenty feet, which I gained by straddling the ridge and working along by my hands. The descent on each side was, as before stated, nearly two thousand feet. I need not say it was the most terrible ride I ever had. Indeed, when I think of it now, it brings up strange and thrilling sensations. How I got over the final peak, I can hardly tell; it seems as if I must have been drunk with excitement, and reached the summit by one of those mysterious chances of fortune which not unfrequently favor men whose minds are in a morbid state.

When I looked down on the waters of the bay, I saw the Brooklyn still at anchor. She looked like some big insect floating on its back, with its legs in the air and little insects running about all over it. I staid up on the top of the Nipple only a few minutes. The view on every side was sublime beyond all the powers of language; but a gust of wind coming, the frail pinnacle of lava upon which I stood swayed, as Abraham had told me; and, fearing it would tumble over, I hurried down the best way I could.

CHAPTER X.

RAMBLE INTO THE INTERIOR.

FINDING by the sun that it was yet early in the day, we resolved, after resting awhile, to push on as far as we could go into the interior. The prospect was perfectly enchanting. Winding ridges and deep gorges lay before us as we looked back from the ocean; and cool glens, shaded with myrtle, and open fields of grass in the soft haze below, and springs bubbling over the rocks with a pleasant music; all varied, all rich and tempting. Away we darted over the rocks, shouting with glee, so irresistible was the feeling of freedom after our dreary ship-life, and so inspiring the freshness of the air and the wondrous beauty of the scenery. The ridge upon which our path lay was barely wide enough for a foothold. It was composed of loose stones and crumbling pieces of clay. The precipice on the right was nearly perpendicular; on the left craggy peaks reared their grizzled heads from masses of dark green shrubbery, like the turrets of ancient castles shaken to ruin by the tempests of ages. Sometimes we had to get down on our hands and knees, and creep over the narrow goat-paths for twenty or thirty feet, holding on by the roots and shrubs that grew in the crevices of the rocks, and at intervals force ourselves through jungles of bushes so closely interwoven that for half an hour we could scarcely gain a hundred yards. About three miles back from the sea-coast, having labored hard to reach a high point overlooking one of the interior valleys, we were stopped by an abrupt rampart of rocks. Here we had to look about us, and consider a long time how we were to get over it.

We now began to suffer all the tortures of thirst after our perilous adventure on the Nipple, and our subsequent struggle through the bushes and along the ridge. There was no sign of a spring any where near; the cliffs were bleached with the wind, and not so much as a drop of water could be found in any of the hollows that had been washed in the rocks by the rain. In this extremity we sat down on a bank of moss, ready to die of thirst, and began to think we would have to return without getting a sight of the valley on the other side of the cliff, when I observed a curious plant close by, nearly covered with great bowl-shaped leaves.

"Abraham," said I, "may be there's water there!"

"May be there is," said Abraham; "let us look."

We jumped up and ran over to where the string plant was, and there we beheld the leaves half full of fine clear water!

"There! what do you think of that, Abraham? Isn't it refreshing? You see it requires a person like me to find fresh water on the top of a mountain where there are no springs."

"Yes, yes," quoth Abraham, slowly, "but may be it's poison."

"Sure enough—may be it is! I didn't think of that," said I, very much startled at the idea of drinking poison. "Suppose you drink some and try. If it doesn't do you any harm, I'll drink some myself in about half an hour."

"Well, I would like a good drink," said Abraham, thoughtfully; "there's no denying that. But it always goes better when I have a friend to join me. I'll tell you what I'll do, Luff. You take one bowl and I'll take another, and we'll sit down here and call it whisky punch, and both drink at the same time."

"Very good," said I, "that's a fair bargain. Come on, Abraham."

So we cut the stems of two large leaves, containing each about a pint of water, and sat down on a rock.

"Your health," said I, raising my bowl; "long life and happiness to you, Abraham!"

"Thank you," said Abraham; "the same to you!"

"Why don't you drink?" I asked, seeing that my friend kept looking at me without touching the contents of the bowl.

"I'm going to drink presently."

"Drink away, then!"

"Here goes!"

But it was not "here goes," for he still kept looking at me without drinking.

"Well," said I, impatiently, "what are you afraid of?"

"I'm not afraid," cried Abraham, "but I don't see you drinking."

"Nonsense, man! I'm waiting for you!"

"Go ahead, then."

"Go ahead."

Here there was a long pause, and we watched each other with great attention. At last, entirely out of patience, I lowered my bowl and said,

"Abraham, do you want me to poison myself?"

"No, I don't," said Abraham; "I'd be very sorry for it."

"Then why did you propose that we should drink this poison together? for I verily believe it must be poison, or it wouldn't look so tempting."

"Because you wanted me to drink it first."

"Did I? Give me your hand, Abraham; I forgot that." Whereupon we shook hands, and agreed to consider it not whisky punch, but poison, and drink none at all.

Our thirst increasing to a painful degree, we were about to retrace our steps, when I observed a little bird perch himself upon the edge of a leaf not far off, and

commence drinking from the hollo· ı ... to look.

"Sure enough," said he, "birds don't drink whisky punch."

"No," said I, "God Almighty never made a bird or a four-legged beast yet that would naturally drink punch or any other kind of poison. It must be water, and good water too, for birds have more sense than men about what they drink. So here goes, whether you join or not."

"And here goes too!" cried Abraham; and we both, without hesitating any longer, emptied our bowls to the bottom; and so pure and delicious was the water that we emptied half a dozen leavesful more, and never felt a bit afraid that it would hurt us; for we knew then that God had made these cups of living green, and filled them with water fresh from the heavens for the good of His creatures.

CHAPTER XI.

THE VALLEY OF ENCHANTMENT.

Thus refreshed, we set to work boldly, and, by dint of hard climbing, reached the top of the cliff. It was the highest point on the island next to the Peak of Yonka. We looked over the edge and down into a lovely valley covered with grass. Wooded ravines sloped into it on every side, and streams wound through it hedged with bushes, and all around us the air was filled with a sweet scent of wild flowers. In that secluded valley, so seldom trodden by the foot of man, we saw how much of beauty lay yet unrevealed upon earth; and our souls were filled with an abiding happiness: for time might dim the mortal eye; the freshness of youth might pass away; all the bright promises of life might leave us in the future; but there was a resting-place there for the

THE VALLEY.

memory; an impression, made by the Divine hand within, that could never fade; a glimpse in our earthly pilgrimage of that promised land where there is harmony without end—beauty without blemish—joy beyond all that man hath conceived.

Nothing was here of that stern and inhospitable character that marked the rock-bound shores of the island.

A soft haze hung over the valley; a happy quiet reigned in the perfumed air; the breath of heaven touched gently the flowers that bloomed upon the sod; all was fresh and fair, and full of romantic beauty. Yet there was life in the repose; abundance within the maze of heights that encircled the dreamy solitude. Fields of wild oats waved with changing colors on the hill-sides; green meadows swept around the bases of the mountains; rich and fragrant shrubs bloomed wherever we looked; fair flowers and running vines hung over the brows of the rocks, crowning them as with a garland; and springs burst out from the cool earth and fell in white mist down into the groves of myrtle below, and were lost in the shade. Nowhere was there a trace of man's intrusion. Wild horses, snuffing the air, dashed out into the valley in all the joyousness of their freedom, flinging back their manes and tossing their heads proudly; and when they beheld us, they started suddenly, and fled up the mountains beyond. Herds of goats ran along the rugged declivities below us, looking scarcely bigger than rabbits; and birds of bright and beautiful plumage flew close around our heads, and lit upon the trees. It was a fair scene, untouched by profaning hands; fair and solitary, and lovely in its solitude as the happy valley of Rasselas.

CHAPTER XII.

A STRANGE DISCOVERY.

While I was trying to make a sketch of this Valley of Enchantment, as we called it, Abraham was peering over the cliff, and looking about in every direction in search of some ruin or relic of habitation. He was not naturally of a romantic turn, but he had a keen eye for every thing strange and out of the way, and an insatiable thirst for the discovery of natural curiosities. Already his pockets were full of roots and pieces of rock; and it was only by the utmost persuasion that I could prevent him from carrying a lump of lava that must have weighed twenty pounds. Without any cause, so far as I could see, he began stamping upon the ground, and then, picking up a big stone, he rolled it over the edge of the cliff, and eagerly peeped after it, holding both hands to his ears as if to listen.

"What's that, Abraham?" said I; "you are certainly losing your wits."

"I knew it! I knew it!" he cried, greatly excited; "it's perfectly hollow. There's a natural castle in it!"

"Where? in your head?"

"No, in the cliff here; it's all hollow—a regular old castle! Come on! come on, Luff! We're bound to explore it. May be we'll rake up something worth seeing yet!" Saying which, he bounded down a narrow ledge on the left, and I, as a matter of course, followed. Our path was not the most secure, winding as it did over an abyss some hundreds of feet in a direct fall; but our previous experience enabled us to spring over the rocks with wonderful agility, and work our way down the more difficult passes in a manner that would have done credit to animals with four legs. Portions of the earth formed a kind of narrow stairway, so distinct and regular that we almost thought it must be of artificial construction. In about ten minutes we reached a broad ledge underneath the brow of the cliff. Turning our backs to the precipice, we saw a spacious cavity in the rocks, shaped a good deal like an immense Gothic doorway, all overhung with vines and wild fern.

"I knew it!" cried Abraham, enthusiastically. "A regular old castle, by all that's wonderful! Crusoe's cave is nothing to it! Just see what a splendid entrance; what ancient turrets; what glorious old walls of solid rock!"

"Verily, it does look like a castle," said I. "We must call it the Castle of Abraham, in honor of the discoverer."

"Yes, but it strikes me there may be another discoverer already. Look at these marks on the rock!"

"True enough; goats never make marks like these!" Near the mouth or entrance of the grotto, traced in black lines, evidently with a burnt stick, we saw a number of curious designs, so defaced by the dripping of water from above that we were unable for some time to make out that they had any meaning. At length, by carefully following the darkest parts, we got some clew to the principal objects intended to be represented, which were very clumsily drawn, as if by an unskillful hand. There was a figure of a man, lying upon a horizontal line, with his face turned upward; the limbs were twisted and broken, and the expression of the features was that of extreme agony; the eyes were closed, the back of the head crushed in, the mouth partly open, and the tongue hanging out. One hand grasped a jagged rock, the other a knife with a part of the blade broken off. Close by, with its head upon his feet, was the skeleton of a strange animal, so rudely sketched that we could hardly tell whether it was intended for a goat or not. It had the horns of a goat, but the eyes, turning upward in their sockets, looked like those of a child that had died some horrible death. Waving lines were drawn some distance off, as representing the sea in a storm; a large ship under sail was standing off in the foam from a pile of rocks that rose out of the sea like a desolate island. The body of a man could be seen under the waves, struggling toward the ship; a shark was tearing the flesh from his legs, and the hands were thrown up wildly over the water. Underneath the whole were several rude sketches of human hearts, pierced through with knives. A hand pointed upward at the figure first described. It had a ring on the forefinger; the tendons of the wrist hung down, as if wrenched from the arm by some instrument of torture. Around these strange designs were numerous others, representing the heads of eagles; a famished wolf, gnawing its own flesh; and the corpses of two children, strangled with a rope; besides other rude sketches of which we could make nothing; and, indeed, some of these already mentioned were so indistinct, that we were forced to depend a good deal on conjecture in order to come to any conclusion in regard to what they were intended to represent; so that I have given but a vague idea, at best, of the whole thing.

"There's something strange about this," said Abraham, trembling all over; "something more than we may like to see. Let us go into the cave, and try if we can solve the mystery."

"I don't think there's much mystery about it," said I; "evidently some sailor who ran away from a ship has occupied this as a hiding-place; these strange designs he has doubtless made in some idle hour, to represent scenes in his own life. The fellow had a bad conscience —he has left the mark of it here."

"He may have left more than that," said Abraham, seriously; "he may have fallen from one of these rocks, and lain here for days, helpless and dying: in the agonies of thirst, driven delirious by fever, he tried, perhaps, to tell by these signs how he died. If I'm not mistaken, we'll find some farther clew to this affair within there. Let us see, at all events."

We then went into the cave, and looked around us as far as the light reached. It was very lofty and spacious, and made a short turn at the back part, so that all beyond was quite wrapt in darkness. Weeds hung in crevices of the dank walls of rock; a few footprints of animals were marked in the ground, some slimy tracks were made over the rocks by snails, and these, together with a dull sound of the flapping of wings made by a number of bats that hung overhead, had a very gloomy effect. However, seeing nothing else in the front part of the cave, we groped our way back into the dark passage at the end, and followed it up till we reached a sort of natural stairway leading into an upper chamber. For some time we hesitated about going up here, thinking there might be a hole or break in the rocks through which by mischance we might fall, and be cast down into some vault or fissure underneath. After a while our eyes got a little used to the darkness, and we thought we could discern the chamber a few steps above into which this stairway led; so we crept up cautiously, feeling our way as we went, and as soon as we found that the ground was level we stood upon our feet, and perceived, from the height above us, and the vacancy all around, that we were in a spacious apartment of the cavern. There still being some danger of falling through, as we discerned by the hollow sound made by our feet, we only went a short distance beyond the entrance, when we stopped still on account of the darkness, which was now quite impenetrable.

"A queer place!" said Abraham; "very like one of the piratical retreats you read about in novels."

"Very, indeed, and quite as unlike reality," said I; "it doesn't seem to be inhabited by pirates now, though, or any thing else except bats. I wish we had a torch, Abraham, for I vow I can't see an inch before me."

"That's not a bad idea," said Abraham; "I think I have a match in my pocket, but it won't do to run the risk of missing fire here. Wait a bit, Luff; I'll go back to the mouth of the cave, and rake up some brush-wood. We'll have some light on the subject presently—if the match don't miss fire."

Abraham then crept back the way we came, as I supposed, for I could see nothing in any direction, and only heard a dull echo around the walls of rock, growing fainter and fainter, till all I was sensible of was the flitting of some bats by my head, and the breath passing through my nostrils. To tell the honest truth, I felt some very queer sensations steal over me upon finding myself all alone in this dark hole, unable to see so much as my hand within an inch of my eyes, and not knowing but the first thing I felt might be a snake or tarentula creeping up my legs, or the bite of some monstrous bat.

I waited with great impatience, without daring to move, lest I should miss the way back and fall through the earth; for in the confusion of my thoughts I had lost all knowledge of the direction of the entrance, and this very thing, perhaps, caused me to magnify the time as it elapsed. It seemed to me that Abraham would never return, he staid away so long, and this brought up some strange and startling thoughts. Suppose, in his search for the brush-wood, he had slipped off the ledge in front of the cave? Suppose he had lost his footing in the dark passage on the way out, and fallen into some unfathomable depth below? Suppose a gang of wild dogs, driven to desperation by hunger, had seized him, and were now, with all their wolfish instincts, tearing him to pieces? The more I thought, the more vague and terrible became my conjectures; till, no longer able to endure the torture of suspense, I shouted his name with all my might. There was no answer but the startling echoes of my own voice, which seemed to mock me in a thousand different directions. I shouted again, and again there was the same fearful reverberation of voices, growing fainter and fainter till they seemed to die upon the air, like the passing away of hope. I now began to peer through the darkness in all directions, with the intention of retracing my steps should I discover any indication of the entrance by which to direct my course. At first it appeared as if the darkness was of the same density all round, but gradually, as I strained my eyes, I thought I perceived a faint glimmer of light, and thither I cautiously made my way, groping about with my hands as I advanced.

In a few moments I felt, by a rush of air, that I was near an opening, and the light growing stronger at the same time, I soon perceived that it led downward in a slanting direction, in the same way as the passage through which we had come up. I was now satisfied that there would be no farther difficulty in getting out, and having no cause to imagine that the place had changed, began to descend as rapidly as possible. All of a sudden my feet slipped from under me, and I went flying down a sort of *chute*, without any power to stop myself, and so terrible was the sensation that I was perfectly speechless, though conscious all the time. It was not long, however, this suspense, for I struck bottom almost at the next moment, and went rolling over headlong into an open space. As soon as I looked around me, I perceived a cleft in the rocks, some fifteen feet above, through which there was a dim ray of light, and this, as I took it, was what had misled me. My sight being rather confused, I now began to grope around me, in order to ascertain if there were any more holes near by, when I discovered that there was straw scattered about over the ground. Instinctively I thought about the strange marks on the rocks near the mouth of the cave. Now if there should be a dead body here, or a skeleton! What a companion in this lonely dungeon! A cold tremor ran through me, and I actually thought that, should I accidentally touch the clammy flesh of a corpse in such a place, it would drive me mad. For a while I scarcely dared to look around, but the absolute necessity of finding some place of exit at last overcame my apprehensions. The light from above was quite faint, as before stated, but yet sufficient, upon getting used to it, to enable me to perceive that I was in a sort of chamber about fifteen feet in diameter, closed on every side except where I had so unexpectedly entered; and I was greatly relieved to find that there was nothing on the ground but a thin layer of straw scattered about here and there, and a few pieces of wood partly burned. I lost no time in making my way into the chute again, which I found but little difficulty in ascending, for it was not so steep as I had supposed. Upon regaining the large apartment from which I had wandered, I heard the muffled echoes of a voice coming, as I thought, from the depths below. They soon grew louder, and I noticed a reddish light faintly shining upon the dark masses of rock. Could it be Abraham? Surely it must be, for I now heard my name distinctly called.

"Halloo there, Luff! Where are you, Luff? Why don't you come on?"

"I'm coming," said I, making a rapid rush toward the light, "as fast as I can."

"All right!" said Abraham; "come on quick!"

It was not long, as may be supposed, before I was scrambling down the rough stairway of rocks by which we had originally entered the mysterious chamber; and the next moment I was standing before Abraham in the passage, which was now no longer dark, for it was lit up with a tremendous torch of brush-wood, which he held in both hands.

"Why, where in the name of sense have you been?" cried he, rather excited, as I thought; "what have you been doing all this time?"

"Doing?" said I; "only exploring the cave, Abraham—hunting up curiosities for pastime."

"Nonsense! I've been calling at you for ten minutes. I didn't want to leave the torch, or I'd have gone up after you; for I couldn't hold it and use my hands at the same time, and I thought if it went out we couldn't light it up again. Besides, I've found a treasure—a treasure, Luff, beyond all price."

"What is it, Abraham—a lump of gold?"

"Pooh! gold couldn't buy it! A skull, sir—a human skull! That's what I've found!"

"Only a skull? I came near finding the whole body," said I, involuntarily shuddering as I thought of the gloomy chamber with the straw in it; "I'm quite certain I'd have found the entire corpse if it had been there."

"But this is a real skull, Luff. It's no subject for trifling. Some poor fellow has left his bones here, as I suspected."

We then went out to the front of the cave. Not far from the entrance was a hole somewhat larger than a man's body, which I had not noticed before, and into which Abraham now crept with the torch, telling me to follow. It was not long before we entered a cell or chamber large enough to stand up in, the floor of which was littered with straw.

"I found it here, Luff; here in this straw—the upper part of a man's skull. Look at it."

Here Abraham removed some of the straw, and there, indeed, lay the frontal part of a skull.

"I found it just as it lies. I put it back exactly in

THE SKULL.

the same position. I wanted you to see how the man died—poor fellow! a sad death he had of it all alone here."

Upon this I took up the skull and examined it. The forehead was small and low, and the whole formation of the upper part of the face somewhat singular. There was not sufficient of the lower part left to tell precisely whether it was the skull of a white man or of a negro. I thought it must be that of a negro, from the size of the animal organs. Abraham, however, considered it the skull of a white man, on account of the whiteness of the bone.

The torch being now burned out, we bethought ourselves of starting toward the valley of the huts, for we had no time to indulge in melancholy reflection on what remained of the poor sailor, or follow up the train of thought suggested by his unhappy fate. Abraham carefully wrapped the skull in his handkerchief, and put it in a large pocket that he had in his coat, declaring, as we set out on our return to the top of the cliff, that a thousand dollars would not induce him to part with so rare and valuable a curiosity.

CHAPTER XIII.

THE STORM AND ESCAPE.

When we reached the summit of the cliff, and looked over once more into the enchanted valley, we could hardly believe that such a change as we beheld could have taken place during our absence. That scene of beauty upon which we had lingered with so much pleasure now seemed to be a moving ocean of clouds, ingulfing every visible point in its billows of mist, raging and foaming as it swelled up over the heights; the wild roar of the tempest vibrating fiercely through the air—the very rocks upon which we stood trembling in the dread coming of its wrath. While we gazed in silence upon the wilderness of surging billows, the whole island became hidden in mist; and that happy valley, so lovely in its solitude but a brief hour before, so calm in its slumbering beauty, so softly steeped in sunshine, was now buried in the fierce conflict of the elements. Nothing was to be seen but an ocean of misty surf below, and a wilderness of dark clouds flying madly overhead. It seemed as if we had been suddenly cut off from the world, and left floating on a huge mass of burned rock, in a chaos of convulsed elements. On every side the impenetrable mists covered the depths, and it needed but a single step to open to us the mysteries of eternity.

The storm set in upon us in fierce and sudden gusts, driving us down for safety upon the lee of the rock. No longer able to stand upright, we cowered beneath the shelter which we found there, and so bided our time. From all we could judge, there was no appearance of a change for the better. As soon as there was a lull, we hurried on along the ridge, in the hope of reaching the valley of the huts before dark, for we had eaten nothing since morning, and were not prepared to spend the night in these wild mountains. After infinite climbing and toil, we came to a part of the path where there were neither trees nor bushes. It was about half a mile in length, and was exposed to the full fury of the gale. About midway we were attacked by a terrific gust of wind and deluge of rain, and it was with great difficulty we could retain our foothold. The rain swashed against us with resistless power, driving us down upon our hands and knees in its fury, while it surged and foamed over us like a white sea in a typhoon. Blinded and dizzy, we rose again and rushed on, staggering in the fierce bursts of the tempest, and gasping for breath in the deluge of spray. How we lived through it I know not; how it was that we were not cast over into the abyss that threatened to devour us, there is but One who knows, for no eye but His was upon us. Breathless, and blinded with the scourging waters, we staggered against a large rock. Here we fell upon our knees, no longer able to contend against the tempest, and clung to the bushes that grew in its clefts, while we silently appealed to Him who holds the winds in the hollow of His hands to take pity upon us, and cast us not away in His wrath.

The worst part of the path being yet before us, where we had previously found it difficult to get over in good weather, we determined upon trying the steep descent on the right, leading directly into the valley of the huts. It was almost a perfect precipice, and was bare and smooth for three hundred yards, where it ran out into a kind of ledge, covered with a stunted growth of trees. If we could reach the grove we would be safe; but between us lay a steep and precipitous field of loose earth, smoothed into a bank of mud by the rains. As we had no alternative, we began the descent as cautiously as possible, thrusting our toes and fingers into the clay, and letting ourselves down by degrees for fifty or a hundred feet at a time, when we stopped a while to look below us. Such was the roar of the storm that I hardly knew whether Abraham was by me or not, when, hearing a loud shout, I looked round and beheld him flying down the precipice with the velocity of lightning. "Oh! he'll be killed!" I exclaimed; "he'll be killed! Oh! what a dreadful death!" At the same moment I felt my hold give way, and I dashed after him in spite of myself, grasping madly at the loose earth, and shouting wildly for somebody to stop me. It was a fearful chase—a chase of life or death! On we sped, upheaving the loose masses of sod, and whizzing through the tempest as we flew; grasping desperately at every rock, tearing up the shrubs that grew in the clefts, and dashing blindly over gaping fissures that lay hidden with the grass. Great masses of burned rock went smoking down into the chaos of mist below, crashing and thundering as they fell. On, and still on, in our wild career we sped, with the vision of death flitting grimly before us! Atoms we were in the strife of elements, whirled powerless into the dark abyss. There was a confused crash of bushes; a stunning sensation—a sudden check—a jarring of the brain—and all was still! I looked, and saw that I was safe. The grove was around me. Consciousness returned as I clung panting to the trees; life was given yet; the vision of death fled in the mists of the tempest.*

* It has already been mentioned that in many parts of the island the soil was loose, and undermined by holes, and the rock weathered almost to rottenness. Pursuing a goat once in one of these dangerous places, the bushy brink of a precipice to which he had followed it crumbled beneath him, and he and the goat fell together from a great height. He lay stunned and senseless at the foot of the rock for a great while—not less than twenty hours, he thought, from the change of position in the sun, but the precise length of time he had no means of ascertaining. When he recovered his senses he found the goat ly-

For a moment, dizzy and confused, I clung to a tree, and offered up my inward thanks to that Providence which had spared me through the fearful ordeal. Then, hearing the voice of Abraham near by to where I stood, I looked, and saw him seated upon the ground, wailing aloud as if in extreme bodily pain. Selfish wretch that I was, had I, in my thankfulness for my own safety, forgotten the friend of my heart! Letting go my grasp of the tree, I ran to his side, and asked in choking accents,

"Abraham! oh, Abraham, are you hurt? Tell me quick—tell me, are you hurt?"

"My skull! my skull!" groaned Abraham, in rending tones; "oh! Luff, my skull is broken!"

"Good heavens!" I exclaimed, "what are we to do? This is terrible! Wretch that I am, I thought only of myself!"

Abraham groaned again. His face was livid, and a small streak of blood that coursed down his right cheek told how truly he had spoken.

"Abraham, my friend Abraham!" I exclaimed, in a perfect agony of distress, "perhaps it's not so bad. It may not be broken."

"Yes it is," said Abraham; "I heard it crack when I fell. My feet flew up, and I fell on my back. It must have struck a rock."

"Oh, Abraham, what are we to do? I wouldn't have had this to happen for the whole island. Here, I'll tear my shirt off and tie it up."

"No, no, Luff, it can't be mended; it's broken all to smash. I wouldn't have had it happen for a thousand dollars. It can never, never be mended!"

"Let me see," said I, carefully laying back his hair; "something must be done, Abraham."

ing dead beside him. With great pain and difficulty he made his way to his hut, which was nearly a mile distant from the spot; and for three days he lay on his bed enduring much suffering. No permanent injury, however, had been done him, and he was soon able to go abroad again.—[LIFE OF ALEXANDER SELKIRK.]

"No, no—nothing can be done; the trouble's not there, Luff; it's *here*—HERE, in my pocket!" At the same time, while I started back in a perfect maze of confusion, Abraham thrust his hand into his coat pocket, and brought forth a whole handful of thin flat bones, broken into small pieces, which he held out with a rueful face, groaning again as he looked at them.

"No, no, it can't be mended, Luff."

"The devil!" said I, angrily, "you may thank your stars it isn't any worse than that!"

"Worse! worse!" cried Abraham, highly excited; "what do you mean? In the name of common sense, isn't that bad enough? How could it be any worse?"

"Pshaw! Abraham; I thought, when I heard your lamentations, and saw that scratch of a bush on your face, that your own natural cranium was fractured."

"Well, what if you did?" cried Abraham, still irritated. "Would you call that worse? A live skull will grow together, but a dead one won't. And this—*this*, with such a history to it—to lose this, after all my trouble in finding it—oh, Luff, Luff, it's too bad!"

However, having no farther time to spare over his ruined skull, he put back the bones in his pocket, and, with a heavy sigh, joined me as I sprang down through the grove.

The rest of our descent was comparatively easy. When we got down to the head of the valley, a muddy stream broke wildly over the rocks, carrying down with it the branches and leaves of trees, and roaring fearfully as it rushed on toward the ocean. We followed this in its rapid descent, and were soon with our friends at the boat-landing.

CHAPTER XIV.

THE AMERICAN CRUSOE.

THE third night closed, leaving us still upon the island. Who could tell if the vessel would be in sight by morning? Should the gale continue, it was not improbable that she would be driven far to the leeward, and perhaps compelled to give up the search for us entirely. Ships had not unfrequently been in sight of the island for weeks, as we afterward learned, and yet unable to make an anchorage, in consequence of baffling winds and heavy gales. It might turn out to be no joke, after all, this wild exhibition. To be Crusoes by inclination was one thing, by compulsion another.

We were determined not to spend another night in the cave; that was out of the question. There was not one of us who wanted to enjoy the romance of that place again. No better alternative remained for us than to make a bargain with Pearce, the American, for quarters in his straw cabin. This we were the more content to do upon seeing him emerge from the bushes with a dead kid hanging over his shoulders, which we naturally supposed he intended for supper.

At first he spoke rather gruffly for a fellow-countryman; but this we attributed to his wild manner of life, separated from all society; nor were we at all disposed to quarrel with him on account of his uncouth address, when we came to consider that a man might understand but little of politeness, and yet be a very good sort of fellow, and understand very well how to cook a kid. We had no money, which we honestly told him in the beginning; but we promised him, in lieu thereof, a large supply of ham and bread from the ship. This did not seem to improve the matter at all; indeed, we began to think he was loth to credit us, which, however, was not

THE AMERICAN CRUSOE.

the case. He said the Californians who had been there had eaten up nearly all his stores, and had paid him little or nothing. They had promised him a good deal, but promises were the principal amount of what he got. If this was all, he wouldn't mind it; they were welcome to what he had; but he didn't like folks to come and take possession of his house as a matter of right, and get drunk in it, and raise Old Scratch with his furniture, and then swear at him next morning for not keepin' a better tavern. He didn't pretend to keep a tavern; it was his own private house, and he wanted it to be private—that's what he came here for. He had society enough at home, and a darn'd sight too much of it. He liked to choose his own company. He was an independent character himself, and meant to be independent in spite of all the Californians on this side of creation. All he wished was that old Nick had a hold of California and all the gold in it—if there was any in it, which he didn't much believe himself. He hoped it would be sunk tolerably deep under the sea before some of 'em got there. It was a tolerable hard case, that a man couldn't live alone without a parcel of fellers, that hadn't any thing to do at home, comin' all the way to Juan Fernandez to play Scratch with his house and furniture, and turn every thing upside down, as if it belonged to 'em, and cuss the hair off'n his head for not makin' a bigger house, and keepin' a bar full of good liquor, and a billiard saloon, and bowlin'-alley for the accommodation of travelers—a tolerable hard case. He'd be squarmed ef he was a goin' to stand it any longer.

We agreed with Crusoe that this was indeed rather a hard case, but promised him that he would find us altogether different sort of persons. We were first-class passengers—none of your rowdy third-class; he understood all that; they were all first-class passengers ashore; he wouldn't believe one of 'em on oath. Again we endeavored to compromise the matter, so far as regarded the ham at least, of which he was entirely incredulous, by telling him that he might come on board with us, and then when we'd be sure not to run away without paying him.

"But what if you should carry ME away?" said he, evidently startled by this proposition.

"Nothing—only we'd take you to California. That would be a lucky chance for you."

"No, it wouldn't. I don't want to go there. I'm very well here."

"But there's plenty of gold in California," said we; "no doubt about it at all. You may live here all your life, and be no better off."

"I'm well enough off," retorted Crusoe; "I only want people to let me alone. Ever since this California business they've been troublin' me."

"You surely can't be happy here without a soul near you! Why, it's enough to drive a man mad. It must be dreadfully dull. You can't be happy!"

"Yes I am!" said Crusoe, peevishly; "I'm always happy when I ain't troubled. When I'm troubled I'm mis'rable. Nothin' makes me so mis'rable as bein' troubled."

"It makes a good many people miserable," was our

reply. "We must trouble you for a night's lodging, at all events, for we have no place else to stay."

"I don't want you to stay nowhere else!" cried Crusoe; "that wasn't what I meant: you mustn't get drunk—that's what I meant."

"No, we won't get drunk; we haven't any thing to get drunk on, unless you insist upon giving us something."

"Very well, then; you can sleep in my cabin, ef you don't tear it down. Some fellers have tried to tear it down."

We promised him that we would use every exertion to overcome any propensity we might have in regard to tearing his house down; and, although he still shook his head mournfully, as if he had no farther confidence in man, he led the way toward his hut, hinting in a sort of undergrowl that it would be greatly to our advantage not to get drunk, or attempt to destroy his house and furniture, inasmuch as he had a number of goatskins, which he wouldn't mind letting sober people have to sleep on, but he'd be squarmed ef he'd lend 'em to people that cuss'd him for not keepin' feather beds. We declared upon our words, as gentlemen, that we had no idea whatever of sleeping on feather beds in such a remote part of the world as this, and would be most happy to prove to him that we were worthy of sleeping on goatskins; that we would regard goatskins in the light of a favor, whereas if he put us upon feather beds, we should feel disposed to look upon it rather as a reflection upon our character as disciples of the immortal Crusoe.

Abraham and myself were wet to the skin after our adventure in the mountains, and, having been five or six hours in that condition, we were hungry enough to eat any thing. We therefore left the party down on the beach, where they were trying to set fire to an old pitch-barrel as a signal for the ship, and, under the guidance of Pearce, hurried up to the cabin. Upon entering the low doorway, we found that there was some promise of good cheer. There was a basket of fish in one corner, and sundry pieces of dried meat hanging upon the walls. Our friend set to work to skin the kid; and we, finding a sort of stone fireplace in the middle of the floor, with a few live embers in it, sat down, and began putting on some wood out of a neighboring pile, by which means we soon had a comfortable fire. As soon as the steam was pretty well out of our clothes, and the warmth struck through to our skins, we felt an uncommonly pleasant glow all over us; and the blaze was exceedingly cheerful. In fact, we were quite happy, in spite of the gloomy forebodings of Pearce, who kept saying to himself all the time he was skinning the kid, "I expect nothin' else but what they'll burn my house down. Ef they'd only let a feller alone, and not come troublin' him, I'd like it a good deal better than bread or ham either—'specially when it's aboard a ship that ain't here, and never will be, I reckon. Fun's fun; but I'll be squarmed ef I want to see my house burned down over my head. 'Tain't nothin' to larf at. When I want somethin' to larf at, I kin raise it myself without troublin' other folks. Ef a man can't live to himself here, I'd like to know where in creation he *kin* live. I expect they'll be explorin' the bottom of the sea by'm-by in search of gold; I'd go there to be to myself, ef I thought I could be to myself; but I know they'd be arter me in less than a month. Ef I was a bettin' character, I'd be willin' to bet five dollars they'll set fire to the house, and burn it down afore they stop!"

Meantime Brigham and the rest of the party succeeded at length in making a large fire on the beach as a signal for the ship, and they remained down there some time in hopes she would send a boat ashore. But the gale increasing, accompanied by heavy rain, they had to leave the fire, and make a hasty retreat to the hut.

CHAPTER XV.

CASTLE OF THE AMERICAN CRUSOE.

PEARCE'S gloomy views of society began to brighten a good deal when he found that we were not disposed to tear down his house or burn it, or wantonly ruin his furniture. He was not a bad-hearted man by any means, though rather crusty from having lived too long alone, and somewhat prejudiced against the Californians on account of the rough treatment he had received from them. A little flattery regarding his skill in architecture, and a word of praise on the subject of his furniture, seemed to mollify him a good deal; and he smiled grimly once or twice at our folly in coming ashore, when we could have done so much better, as he alleged, by staying aboard the ship, and going ahead about our business.

Regarding the house, which afforded him so much anxiety, there did not appear to us to be any thing quite so original and Crusoe-like in any other part of the world. It was a little straw hut, just big enough to creep into and turn round in; with a steep peaked roof, projecting all round, very rustic and rugged-looking, and, withal, very well adapted to the climate. The straw was woven through upright stakes, and made a tolerably secure wall; outside, growing up around the house in every direction, were running vines and wild flowers; and at a little distance were various smaller sheds and out-houses, in which our worthy host kept his domestic animals, and what wood he required during the bad weather. The furniture of his main abode, which was such a source of honest pride to him, consisted chiefly of a few three-legged stools, made of the rough wood with the bark still on; a kind of bench for a lounge; a rough bedstead in one corner, partly shut off by a straw partition; a broken looking-glass, and an iron kettle and frying-pan, besides sundry strange articles of domestic economy of which we could form no correct idea, inasmuch as they were made upon novel principles of his own, and were entirely beyond our comprehension. Over head, the rafters were covered with goatskins; a sailor's pea-jacket, a sou'wester, and some colored shirts hung at the head of the bed. In one corner there was a rude wooden cupboard, containing a few broken cups and plates, and a Chinese tea-box; in another a sea-chest, which, when pulled out, served for a table. The floor was of mud, and not very dry after the rain; for the roof had sprung a leak, and, moreover, what water was cast off from above eventually found its way in under the walls below. Doubtless, like the man with the fiddle, our host thought it useless to mend it when the weather was fine, and too wet to work at it when the weather was rainy. It was a very queer and original place altogether; and with a good fire, and a little precaution in keeping from under the leaks in the roof, not at all uncomfortable. Our Crusoe friend, overhearing us say that it was a glorious place to live in, a regular castle, where a man might spend his days like a king, smiled again a crusty smile, and growled,

"There's tea in that 'ere box. Ef you want some you kin have it. I got it out'n a ship that came from China. There ain't better tea nowhere."

We thanked him heartily for his kindness, and declared at the same time that we regarded good tea as the very rarest luxury of life. Again his face cracked into something like a smile, and he said,

"Better tea never was drunk in China. Ef you like, I'll put sugar in it."

We declared that sugar was the very thing of all the luxuries in the world that we were most attached to, but we could not drink it with any sort of relish if we thought it would be robbing him of his stores. If he had these things to spare we would cheerfully use them, and pay him three or four times their value in provisions from the ship.

"Darn the ship!" cried Crusoe; "I don't care a cuss about the ship, so long as you don't get drunk and tear my house down!"

Upon this we protested that we would sooner tear the hair out of our heads by the roots than tear down so unique and extraordinary a structure as his house; and as to his furniture, it was worth its weight in gold; every stick of it would bring five hundred dollars in the city of New York.

Whereupon Pearce stirred about in the obscure corners with wonderful alacrity, rooting up all sorts of queer things out of dark places, and muttering to himself meantime,

"I'm as fond of company as any body, ef they're the right sort; and I'll be squarmed ef I ain't an independent character too. I don't owe nobody for a buildin' of my house, or a makin' of my furniture. I did it all myself, long before California was skeer'd up."

He then put down the old kettle on the fire, and, as soon as the water was boiled, emptied a large cupful of tea into it, and set it near the fire to draw. While the tea was drawing, he fried a panful of kid, and broiled some fish on the coals; and when it was all done, he gave us each a tin plate, and told us to eat as much as we wanted, and be darn'd to the ship, so long as we behaved like Christians. Then he furnished us with cups for the tea, and some sea-biscuit, which he dug out of the cupboard; and I must declare, in all sincerity, that we made a most excellent supper.

CHAPTER XVI.

DIFFICULTY BETWEEN ABRAHAM AND THE DOUBTER.

EVERY one of us, except the man that had no faith in Robinson Crusoe, admitted that the tea was the best ever produced in China or any where else; that the fried kid was perfectly delicious; that the fish were the fattest and tenderest ever fished out of the sea; that the biscuit tasted a thousand times better than the biscuit we had on board ship; that the whole house and all about it were wonderfully well arranged for comfort; and that Pearce, after all, was the jolliest old brick of a Crusoe ever found upon a desolate island.

In fine, we came to the conclusion that it was a glorious life, calculated to enlarge a man's soul; an independent life; a perfect Utopia in its way. "Let us," said we, "spend the remainder of our days here! Who cares about the gold of Ophir, when he can live like a king on this island, and be richer and happier than Solomon in his temple!"

"You'd soon be tired of it," muttered a voice from a dark corner: it was the voice of the Doubter. "You wouldn't be here a month till you'd give the eyes out of your heads to get away."

"Where's that man?" cried several of us, fiercely.

"I'm here—here in the corner, gentlemen, rayther troubled with fleas."

"You'd better turn in and go to sleep."

"I can't sleep. Nobody can sleep here. I've tried it long enough. I reckon the fleas will eat us all up by morning, and leave nothing but the hair of our heads. I doubt if they'll leave that."

"Was there ever such a man? Why, you do nothing but throw cold water on every body."

"No I don't; it comes through the roof. It's as much as I can do to keep clear of it myself, without throwin' it on other people." With this we let him alone.

The fire now blazed cheerfully, sending its ruddy glow through the cabin. A rude earthen lamp, that hung from one of the rafters, also shed its cheerful light upon us as we sat in a circle round the crackling fagots; and altogether our rustic quarters looked very lively and pleasant. Every face beamed with good-humor. Even the face of the Doubter belied his croaking remarks, and glowed with unwonted enthusiasm. Little Jim Paxton, the whaler, under the inspiration of the tea, which was uncommonly strong, volunteered a song; and the cries of bravo being general, he gave us, in true sailor style,

"I'm monarch of all I survey,
My right there is none to dispute;
From the centre all round to the sea,
I'm lord of the fowl and the brute!
Oh Solitude where are the charms," &c.

This was so enthusiastically applauded, that my friend Abraham, whose passion for all sorts of curiosities had led him to explore musty old books as well as musty old caves for odds and ends, now rose on his goatskin, and said that, with permission of the company, he would attempt something which he considered peculiarly appropriate to the occasion. He was not much of a singer, but he hoped the interest attached to the words would be a sufficient compensation for all the deficiencies of voice and style.

"Go ahead, Abraham!" cried every body, greatly interested by these remarks. "Let us have the song! Out with it!"

"First," said Abraham, clearing his voice, "I beg leave to state, for the benefit of all who may not be familiar with the fact, that this is no vulgar or commonplace song, as many people suppose who sing it. On the contrary, it may be regarded as a classical production. Among the many effusions to which the popularity of

Robinson Crusoe gave rise, none was a greater favorite in its day than the song which I am about to attempt. It has been customary to introduce it in the character of Jerry Sneak, in Foote's celebrated farce, the Mayor of Garratt. As the words are now nearly forgotten, I hope you'll not consider it tiresome if I go through to the end. Join in the chorus, gentlemen!"

POOR ROBINSON CRUSOE.

"When I was a lad, my fortune was bad,
My grandfather I did lose O;
I'll bet you a can, you've heard of the man,
His name it was Robinson Crusoe.
Oh! poor Robinson Crusoe,
Tinky ting tang, tinky ting tang,
Oh! poor Robinson Crusoe.

"You've read in a book of a voyage he took,
While the raging whirlwinds blew, so
That the ship with a shock fell plump on a rock,
Near drowning poor Robinson Crusoe.
Oh! poor, &c.

"Poor soul! none but he escaped on the sea.
Ah, Fate! Fate! how could you do so?
'Till at length he was thrown on an island unknown,
Which received poor Robinson Crusoe."

"Here, gentlemen, I beg you to take notice that we are now, in all probability, on the very spot. I have the strongest reasons for supposing that the castle of our excellent host, in which we are at this moment enjoying the flow of soul and the feast of reason, is built upon the identical site occupied in former times by the castle of the remarkable adventurer in whose honor this song was composed. But to proceed—

"Tinky ting tang, tinky ting tang,
Oh! poor Robinson Crusoe.

"But he saved from on board a gun and a sword,
And another old matter or two, so
That by dint of his thrift, he managed to shift
Pretty well, for poor Robinson Crusoe.
Oh! poor, &c.

"He wanted something to eat, and couldn't get meat,
The cattle away from him flew, so
That but for his gun he'd been sorely undone,
And starved would poor Robinson Crusoe.
Oh! poor, &c.

"And he happened to save from the merciless wave
A poor parrot, I assure you 'tis true, so
That when he came home, from a wearisome roam,
Used to cry out, Poor Robinson Crusoe.
Oh! poor, &c.

"Then he got all the wood that ever he could,
And stuck it together with glue, so
That he made him a hut, in which he might put
The carcass of Robinson Crusoe."

"Hold on there! hold on!" cried a voice, in a high state of excitement. Every body turned to see who it was that dared to interrupt so inspiring a song. Immediately the indignant gaze was fixed upon the face of the Doubter, who, with outstretched neck, was peering at Abraham from his dark corner. "Excuse me, gentlemen," said he, "but I want some information on that point. Did you mean to say, sir, that he, Robinson Crusoe, stuck the wood together with *glue* when he built his house? with GLUE, did you say?"

"So the song goes," said Abraham, a little confused, not to say irritated. "Doubtless the words are used in a metaphorical sense. There is every reason to believe that this is a mere poetical license; but it doesn't alter the general accuracy of the history. For my own part, I am disposed to think that the house was built very much upon the same principles as that of our friend Pearce; in fact, that it was precisely such an establishment as we at present occupy."

"Go on, sir—go on; I'm perfectly satisfied," muttered the Doubter; "the whole thing hangs together by means of glue; every part of it is connected with the same material!"

Abraham reddened to the eyebrows at this uncalled-for remark; his fine features, usually so placid and full of good nature, were distorted with indignation; he turned fiercely toward the Doubter; he instinctively doubled up both fists; he breathed hard between his clenched teeth; then, hearing a low murmur of dissuasion from the whole party, he turned away with a smile of contempt, breaking abruptly into the burden of his song,

"Tinky ting tang, tinky ting tang,
Oh! poor Robinson Crusoe!

"While his man Friday kept the house snug and tidy,
For be sure 'twas his business to do so,
They lived friendly together, less like servant than neighbor,
Lived Friday and Robinson Crusoe.
Oh! poor, &c.

"Then he wore a large cap, and a coat without nap,
And a beard as long as a Jew, so
That, by all that's civil, he looked like a devil
More than poor Robinson Crusoe."

"Which shows," continued Abraham, with his accustomed smile of good humor, "the extraordinary shifts to which a man may be reduced by necessity, and the uncouth appearance he must present in a perfectly unshaved state, when even the poet admits that he looked like a devil. These articles of clothing, which contributed to give him such a wild aspect, were made of goatskins, as he himself informs us in his wonderful narrative; and I beg you to remember, gentlemen, that the very skins upon which we are this moment sitting are related, by direct descent, to those which were worn by Robinson Crusoe."

Here the Doubter groaned.

"Well, sir, is there any thing improbable in that?" said Abraham, fiercely. "Have you any objection to that remark, sir?"

"No; I have nothing to say against it in particular, except that I'd believe it sooner if there were goats in the skins. I never heard of modern goatskins descending from ancient goatskins before."

"Of course, sir," said Abraham, coloring, "the goats were in the skins before they were taken out."

"Likely they were," growled the Doubter; "I won't dispute that. But I'd like to know, as a matter of information, if he, Robinson Crusoe, made his clothes in the same way as he made his house?"

"To be sure, sir; to be sure: he made both with his own hands."

"I thought so," said the Doubter, sinking back into his dark corner; "he sew'd 'em with glue. All glue—glue from beginning to end."

"I'll see you to-morrow, sir!" said Abraham, swelling with indignation; "we'll settle this matter to-morrow, sir. At present I shall pay no further attention to your remarks!" Here he drew several rapid breaths, as if swallowing down his passion; and, looking round with a darkened brow upon the mute and astonished company, resumed, in a loud and steady voice,

"Tinky ting tang, tinky ting tang,
Oh! poor Robinson Crusoe!

"At length, within hail, he saw a stout sail,
And he took to his little canoe; so,
When he reach'd the ship, they gave him a trip,
Back to England brought Robinson Crusoe.
Oh! poor Robinson Crusoe!"

We all joined in the chorus—all, except the incredulous man; and, notwithstanding the unfortunate difference between Abraham and that individual, which tended so much to mar the harmony of the occasion, we thought, from the way our voices sounded, that it must have been the very first time this inspiring song was sung in the solitudes of Juan Fernandez. I even fancied I detected the crusty voice of Pearce in the chorus; but I wouldn't like to make a positive assertion to that effect, on account of the danger of giving him offense, should he ever cast his eyes upon this narrative. As there was still evidently a cloud upon Abraham's brow, which might burst to-morrow upon the Doubter, and thereby bring the whole adventure to a tragic termination, several of us now, by a concerted movement, endeavored to effect a reconciliation. We seized upon the Doubter, who by this time was dozing away in the corner, and brought him forth to the light, where he looked about him in mute astonishment, muttering, as if awakened out of a dream, "No, sir, it can't be done, sir; a house never was built with glue yet; goatskins never were sewed together with glue—never, sir, never!"

"You shall swallow those words, sir!" cried Abraham, quivering with passion; "I'll make you swallow them, sir, to-morrow morning!"

"I'll swallow 'em now if you like," drawled the Doubter, with provoking coolness, "but I can't swallow a house built of glue. Possibly I might swallow the goatskins, but the house won't go down—it ain't the kind of thing to go down!"

Here it required our full force to restrain Abraham; he fairly chafed with indignation; his face was flushed; his nostrils distended; his stalwart limbs writhing convulsively; in truth, our well-meant plan of reconciliation only seemed to hasten the tragedy which we were striving to prevent. Pearce himself now interposed.

"I know'd it," said he; "I know'd they'd tear my house down yet, and ruin my furniture! Next thing, all hands'll be breakin' my chairs to pieces on one another's heads; I know'd it; I wouldn't believe 'em on oath!"

This rebuke touched Abraham in a tender point. Quick to take offense, he was also ready in forgiving an injury, especially when a due regard for the feelings of others required it.

"Gentlemen," said he, "it shall never be said that I have violated the rites of hospitality. There shall be no further difficulty about this matter; I forgive all. Your hand, sir!"

The Doubter awkwardly held out his hand and suffered it to be shaken, upon which he crept back into his dark corner, still, however, muttering incoherently from time to time; but as nothing could be distinguished but the word "glue," it was not deemed of sufficient importance for the renewal of hostilities, or the interruption of the general harmony. Good humor being restored, it was all the more hearty after these unpleasant little episodes; and so genial an effect had it upon Pearce, that he quite forgot his resentment, and unbended himself again. Gradually he began to tell us wild stories of his Crusoe life; how he had lived all alone for nearly a year on the island of Massafuero without seeing the face of man; how, during that time, he sustained himself upon roots and herbs, and likewise by catching wild goats in traps; how he never was so happy in his life, and never had any trouble till he left that island in a whaler, and came here to Juan Fernandez; how for two years he had lived on this island, sometimes alone, and sometimes surrounded by outlawed Chilians; how on one occasion, while up in the mountains hunting goats, he fell down a precipice, and broke his arm and two of his ribs, and was near dying all alone, without a soul to care for him. A great many strange stories and legends he told us, too, in his rude way, about Juan Fernandez; and so strong was his homely language, and so fresh and novel his reminiscences, that we often looked round in the waning light of the lamp for fear some ghost or murderer would steal in upon us.

As well as I can remember, one of his strange narratives was substantially as follows. There was all the force of reality to give it interest; for it was evidently, as he told us, a simple recital of facts.

CHAPTER XVII.

THE MURDER.

About five years ago (I think he said it was in 1844), a murder was committed on the island by the father of one of the present Chilian residents. Pearce was then in Valparaiso, and had a statement of the circumstances from some of the parties concerned in it.

TRAGIC FATE OF THE SCOTCHMAN.

A Scotch sailor, it appeared, deserted from a vessel that touched at the island for wood and water. For a time he concealed himself in a cave among the cliffs near the bay. When the vessel sailed, he came down into the valley and built himself a hut out of straw, in which he resided several months alone. By fishing, and catching wild goats in traps, he supported himself comfortably, and was becoming reconciled to his isolated life, when a family of Chilians, consisting of five or six men and women, under the control of an old Spaniard, father-in-law of one of the younger men, came over about this period in a small trading vessel from Massafuero. They had been living there for some time, but thought they could do better in Juan Fernandez. There were no huts standing there then except that belonging to the sailor. The Chilians prevailed upon him to let them occupy a part of his house, promising to build themselves one as soon as they could cut straw and wood enough. Every day they went out on the hill-sides to cut the straw, and they seemed to be making good progress with their hut. One night the sailor, as he lay in bed, overheard one of the Chilians say to the others, "We are working hard every day, but it will be a long time before we can get a house built. Neither will it be big enough for us all when we finish it. This man is nothing but a heretic, therefore it would be no sin to take his life. Let us kill him, and then we can have his house, which has other buildings to it, without the trouble of doing any more work." The others agreed to this, all except one woman, who said God would never suffer them to prosper if they committed such a deed. However, they silenced her by threats, and then talked further upon the best means of murdering the Scotchman. Having been a beach-comber for many years in Spanish countries, he understood the language, and it so happened that he overheard nearly every word. Being a powerful man, of great courage and fierce temper, he sprang from his bed, and swore they must leave the house at that very instant, or he would cut their throats. The woman he would have spared this treatment, but he knew she would only fare the worse for his protection. Finding him resolute, they took their things and left the house; but after they were out in the dark, it being a stormy night, they begged so hard for shelter that he told them they might go into a shed, which he had built some distance off to keep goats in. Here they remained, without daring to molest him, until their own house was completed. In the mean time, the suspicions of the sailor were lulled by their friendly behavior, and he often spent a part of his time in social talk with them, which was the more agreeable inasmuch as the old man's daughter, who had taken his part at first, fell in love with him, and, although jealously watched by her husband, found frequent chances of meeting him alone. He became much attached to her, as well on account of her attempt to save his life as the charms of her person, which were well calculated to excite admiration and kindle the amorous flame. She was a very beautiful woman, a Chilian by birth, and was married against her inclination; and coming from a country where the marriage tie is not considered so sacred as it is in more northern climes, she had but little scruple in yielding to her guilty love. His manly person and bold bearing had attracted her in the first place, and these stolen interviews only served to strengthen the passion that grew up between them. At this period they were joined by an English sailor,

THE LOVERS.

another deserter, who took up his quarters with the Chilians in their new abode, and became a member of their gang. The Scotchman had refused, from some dislike that he formed to this man on first sight, to take him into his cabin. This led to a mutual hatred, which was soon increased by other causes. The Englishman, struck by the beauty of the young woman, whose affections the other had won, now made love to her on all occasions, but she gave him no encouragement. He attributed his failure to the Scotchman, whom he secretly watched. Fired with jealousy and deadly hatred toward his rival, he resolved upon putting him to death by stratagem, for he was too cowardly to undertake it openly. Having learned the difficulty that had previously occurred, he took occasion to tell the Chilians that the Scotchman was their mortal enemy, and only awaited an opportunity to murder them all, so as to get entire possession of the young woman, with whom he had already formed a guilty connection. At this period three Americans deserted from a whale-ship and joined the Scotchman. Through some accident, or most likely by foul means, his hut took fire soon after, and was burnt to the ground. He and his companions were obliged to move to a cave near by, where they designed living till they could build another. Knowing nothing of the schemes of the English sailor, who took care that it should not be found out through the woman, they were ignorant of the hostile intention of the Chilians, till one day, as they were scattered over the valley, cutting wild oats for their cabin, the Englishman told the old man, who was the leader of the Chilians, that he had overheard the other party say they were going to murder them all that night; and prevailed upon him to muster his men together secretly, and settle the matter at once. They all went first to the cave, and took possession of the arms left there by the Americans and their leader. The old man, followed at a distance by his comrades, thereupon proceeded to the valley with a loaded gun; and seeing the Scotchman at a distance from the others, he stole upon him and shot him through the body with slugs. Badly wounded, but not mortally, the Scotchman shouted to his friends that he was shot; that they must follow him and fight for their lives, upon which he ran, covered with blood, toward the cave, followed by the Americans. On arriving there they found all their fire-arms gone: they fought for some time with their knives, but were finally overpowered by the Chilian party and bound hand and foot.

Next day it so happened that a whale-ship came into the harbor for wood and water. The Americans were carried back some distance and hid among the cliffs, with an armed guard over them, so that they might be out of the way when the people from the ship came ashore; and the wounded man was concealed in a cave. The

Englishman then went on board with the old Chilian, and told the captain that a deserter from a whale-ship, who had been on the island some time, had undertaken to murder them, and they had shot him in self-defense. Their story was plausibly told, and was believed. They said the man was not dead, and they asked the captain to take him away, as they wanted to get rid of him. The captain refused to do this, saying he would have nothing to do with a deserter; if the man got into trouble by his misconduct, he might get out of it the best way he could. When the vessel sailed, which was the next day, the Chilians, in compliance with the advice of the Englishman, took their wounded prisoner out into an open space, and shot him through the heart. He fell dead upon the spot. They then dug a hole in the ground and buried him; and, in order to keep his spirit from rising upon them at night, they erected a cross over the grave. The woman, upon hearing that her lover was

GRAVE OF THE MURDERED MAN.

murdered, fell into a state of melancholy, and refused to taste any food for many days. Such was her distress, that she wandered about the cliffs like one bereft of her senses, and was often found at night weeping upon his grave. Indeed, she never fully recovered, but was always from that time weakly and unsettled in her mind.

Another vessel came into port in the course of a few months, and the affair became known through the three Americans, who made their escape and got on board. News of the murder was carried to Talcahuana by this vessel; and as soon as it reached Valparaiso, a small Chilian cutter, then lying in the harbor, was dispatched to the island of Juan Fernandez to capture and bring home the murderers. On their arrival in Valparaiso they were taken in irons to Santiago, the seat of government, where they were tried and sentenced to be shot in the public plaza. Some of the circumstances, considered palliating, became known before the execution was carried into effect, and their punishment was commuted to five years' banishment on the island of St. Felix.

The Chilian government still holds a penal settlement on that island. All criminals of a desperate character are sent there and subjected to hard labor. The term for which these murderers had been banished had just expired (in 1849), and it was supposed by the present Chilian residents that they would return by the first opportunity to Juan Fernandez.

CHAPTER XVIII.

THE SKULL.

During the recital of this tragical narrative, Abraham, who had listened to every word with intense interest, became strangely agitated. Several times it was apparently with the utmost difficulty he could refrain from relieving himself of something that produced an unusual effect upon his mind. Especially when it came to the death of the unfortunate Scotchman, I thought I noticed that he was intensely excited. At first, knowing the tenderness of his feelings, I attributed this extraordinary manifestation of interest to grief and pity for the unhappy fate of the beautiful Chilian; but I soon found that it proceeded from another and very different cause. No sooner had Pearce concluded than he exclaimed,

"I'll wager a thousand dollars, gentlemen, that the Scotchman never was buried!"

"He was buried, certain," said Pearce; "I can show you the place."

"Then there is some strange mystery about it," said Abraham, somewhat disappointed. "This very day I found a man's skull, which I am now quite certain has some connection with this tragedy."

The intense excitement produced by this disclosure is quite indescribable. Every body in the party leaned forward, with starting eyes, and gazed with breathless interest at Abraham. He had purposely withheld making any reference to the affair of the skull till a fitting opportunity should occur to disclose all the particulars, when the mind of every individual present was in a proper tone of solemnity to receive so important a communication. That opportunity had now occurred, under the most favorable and unlooked-for circumstances. I never saw Abraham so excited in my life before—not even on the occasion of his late unpleasant difficulty.

"Gentlemen," said he, "I had a presentiment before we left the ship that this expedition would result in some extraordinary discovery. You may judge from the facts which I am about to disclose to you how far this presentiment has been verified."

He then, in a voice of becoming solemnity, went into a detailed narrative of our adventures in the mountains. He commenced at the very starting-point, where we separated from the hunting party; he dwelt vividly on our perilous adventure on the cliff, stating all the particulars of our escape; how we climbed up a perpendicular wall of rocks four thousand feet high; how we stood upon the very highest pinnacle, which was only ten inches in diameter; how, when we came down again to the base, we lay perfectly insensible for an entire hour;

and the wonderful adventures we had in the interior—the walk of six miles directly back from the ocean; our preservation from a horrible and lingering death by thirst, through the agency of a little bird; the Enchanted Valley that we explored, and the two wild horses we caught entangled in the bushes, and afterward rode; our discovery of an old castle built in the sixteenth century by Juan Fernando; the mysterious marks upon the outer wall; our strange and startling explorations of the interior vaults and marble halls; and finally the discovery of the skull—the skull of some unfortunate man who had crept into one of those dreary vaults, where he died on a miserable bed of straw, all alone, without a soul near him! Afterward how he (Abraham) and myself were overtaken by a frightful tornado, and cast down over the rocks a distance of three miles in a direct line; how, during this terrible fall, he had the misfortune to strike a rock, and ruin the invaluable relic of mortality which he had put in his pocket, by breaking it all to pieces; but—

"Did you save the pieces?" asked a voice from the corner. Of course it was the voice of the Doubter. A look from Abraham silenced him, and the narrative was resumed:

But it fortunately happened that a portion of the socket of one eye and a piece of the forehead remained entire, which, together with all the smaller fragments, he would be most happy to exhibit to the company; premising, however, that there was but little question in his mind, from all the particulars of Pearce's tragical narrative, that this skull was in some way or other connected with it. Possibly it might be that the unhappy young woman, who it appears was the victim of an inordinate passion for the murdered man, bereft of her senses by his tragical death, went to his grave at night and dug up his body, and being unable to carry it away at once, perhaps she cut it to pieces, and carried it by degrees up to her secret place of wailing in the mountains, where she could mourn over his remains without fear of discovery. It was not an unreasonable conjecture, he thought, considering the woman was insane. In some hour of despondency she had probably made those mysterious designs which had led to the discovery—the sketch of the dead body of her lover; the ship that left the island without saving him; some pet goat that doubtless accompanied her in her wanderings; the children that were strangled, and all those vague marks, which indicated the character of her thoughts.

During the narration of these adventures, which I must confess astonished me not a little, well as I knew the enthusiastic character of my friend (and he never was more in earnest in his life), I observed that Pearce had doubled himself up almost into a knot, covering his face with his hands, and heaving convulsively, as if moved by some internal earthquake. There was no sound escaped him, but it was quite evident that he was strangely affected by Abraham's narrative. The rest of the party were so deeply interested in the whole disclosure that they took no notice of him. Could it be that Pearce himself was implicated in the murder? That it was all a fiction his being in Valparaiso at the time? That he was in any way attached to this unfortunate female, whose sad fate had aroused all our sympathies?

"I'd like to see that skull," said the Doubter.

"Here it is—or what remains of it," said Abraham, drawing forth the pieces from his pocket; "you can all see it if you wish."

The pieces were handed round and examined with intense interest and curiosity.

"You call this a man's skull?" said the Doubter, looking incredulously at a piece which he held in his hand.

"I do, sir," said Abraham, sharply; "have you any objection to my calling it a man's skull, sir?"

"No, none at all; you may call it a dog's skull if you like. *I'd* call it Robinson Crusoe's skull if I owned it. For all I know to the contrary, it *is* his; but I'd like to have a certificate from himself to that effect before I'd place much confidence in my own opinion, if I thought so."

The biting satire of these remarks touched Abraham to the quick. Nothing in the world would have prevented him from springing upon the Doubter at that moment, and taking summary vengeance upon his person, but the sudden exit of Pearce, who, rising from his goatskin, hurriedly left the cabin. This produced a general murmur of disapprobation. It was the unanimous opinion that a course of conduct, resulting as this did—compelling a man, as it were, to leave his own castle for personal security, was very unbecoming; and that Abraham, being the chief, although perhaps unintentional cause of it, was in honor bound to go after him and bring him back.

I take pride in saying that my friend was not the kind of man to resist such an appeal as this. He immediately left the hut and went in search of Pearce. Meanwhile we took occasion to administer a well-merited rebuke to the Doubter; and to declare that if he again interrupted the harmony of the evening, we would leave him ashore when we started for the ship. His only reply to this was, that he hoped, if he should unfortunately die in a cave in consequence of our cruelty, that his head would make a better-looking skull than the one Abraham had found.

In about ten minutes Abraham and Pearce returned, both having a very strange expression upon their features. Pearce looked unnaturally serious about the mouth, but I fancied more knowing than usual about the eyes. In sitting down he dropped a dollar, which he hastily picked up and put in his pocket. As to my friend, I thought there was something confused and dejected in his look; but he immediately said with assumed spirits when he came in, "All right, gentlemen; all right. The whole thing is settled; let there be nothing more about it."

Some few questions, however, were asked concerning the skull, but all the satisfaction Abraham could give was, "You have the particulars, gentlemen; you must judge for yourselves." Pearce professed to know nothing about it.

Harmony and good-humor being again restored, there were numerous calls for some farther reminiscences of the island.

Pearce said he didn't know whether any of us had ever heard of the governor's vision; if we hadn't, maybe we'd like to hear something about it. He couldn't promise that it was all true, but the Chilians here believed every word of it; "and, likely enough," he added, looking quietly at Abraham, "there may be some of you that can account for it."

"Let us have it!" exclaimed every body in a breath; "the governor's vision, by all means."

Pearce then fixed himself comfortably on his goatskin, and, putting some fagots on the fire, gave us in substance the following history of

THE GOVERNOR'S VISION.

CHAPTER XIX.

THE GOVERNOR'S VISION.

THE highest peak on the island of Juan Fernandez is called the Peak of Yonka. It forms an abrupt precipice all round, of several hundred feet. Various attempts have been made from time to time, by sailors and others, to ascend to the summit, but this feat has never been accomplished except in a single instance. A cross still stands upon it, which was erected by two Chilians many years ago, under very strange circumstances. It appears that the Chilian governor at the time of the penal settlement in Cumberland Bay went out riding one day near this mountain. On his return he related to his people a strange vision which he had seen in the course of his ramble. He said that, while looking at the peak, he saw down in the valley that lay between a tall man dressed in black, with a black hat on, mounted on a horse of the purest white. The strange rider turned toward him, showing a face of ghastly paleness. He looked at him steadfastly, with "eyes of fire," as the governor declared, the glare of which made the air hot all around. The governor, trembling with awe, made the sign of the cross, upon which the strange horseman put spurs to his horse, and rode straight up the precipice to the summit of the peak, where he stopped a moment to look back. He then, upon seeing the sign of the cross made again, waved his hands wildly, as if in despair, and plunged out of sight on the other side. Being a devout man, and withal a believer in spirits, the governor considered this to be an omen of some impending calamity, which could only be averted by planting a cross on the peak. For this purpose he selected two criminals, under sentence of death for the murder of a soldier, and offered them their liberty if they would make the ascent and erect the cross. In the one case there was the certainty of death, in the other a chance of life. The criminals therefore resolved to make the attempt. Ropes, ladders, and tools were furnished them by the governor, and they were allowed such provisions as they required, with injunctions that at the expiration of ten days, in case of failure, they would be executed. For eight days they toiled incessantly. They drove spikes into the walls of rock, and day by day went up a little higher, letting themselves down again at night by ropes to the base of the precipice. On the eighth day they reached the summit, ready to die of fright, and worn to skeletons at the terrible ordeal through which they had passed. It took them all the next day to recover sufficiently to be able to resume their labors. The table on the top was of solid rock, not more than fifty feet in diameter. In the centre was a spring of clear water bubbling up and running over the rocks. One of them bathed in this water, and was so refreshed that he thought it must have some magical properties. He went over to the edge on the western side, and looked down to see where it fell. Directly beneath him, he saw a line stretched from two points of rock over the precipice, nearly covered with linen shirts, as white as the driven snow, and apparently of the finest texture. He called to his comrade to come and witness this wonderful sight. While the two men were looking over, there came a tremendous hurricane, that compelled them to throw themselves flat on their backs to avoid being blown over into the abyss. After the hurricane had passed they again looked over, but the line and shirts had disappeared, and they saw nothing but the bare rocks. They then fell upon their knees and prayed, and the vision of an angel appeared to them, telling them to put up the cross near the spring. As soon as they had planted the cross, they let themselves down by the ropes, and hastened to tell the governor of the strange adventures that had befallen them. So impressed was he by their wonderful narrative, that he immediately gave them their freedom, as he had promised, and sent them home laden with presents; and he had crosses erected on various parts of the island, and masses performed by the soldiers for a long time after.

"I wouldn't swear to it all," added Pearce, looking again toward Abraham. "But likely some of you gentlemen, who have more schoolin' than I have, may be able to account for it."

Abraham reddened a little and looked confused, but said nothing. A voice from the corner broke in,

"I know exactly how it happened; nothing is easier than to account for it. In the first place, it didn't happen at all. The governor was dyspeptic. I'm raything dyspeptic myself, gentlemen, and I know what sights a man sees when he gets the horrors from dyspepsia. I've seen stranger sights than that when it was bad on me—once, in particular, I was troubled a good deal worse than the governor."

"Impossible," said Abraham, scornfully, "utterly impossible, sir, that you could ever have seen any thing half so strange as the governor's vision."

"I didn't see a house made of glue," retorted the Doubter. "I didn't ride on wild horses; neither did I find a castle with a skull in it. I didn't carry the skull six miles, and then find out that it came off the head of a four-legged man; and that the four-legged man was cut to pieces by his lady-love; but I'll tell you what I did see."

"Hold, sir, hold!" cried Abraham, now perfectly furious. "By heavens, gentlemen, I can't stand such insults as these! You must suffer me to chastise this wretch. Miserable poltroon! do you dare to taunt me in that manner? I'll see you, sir—I'll see you to-morrow morning!"

"Likely you will," said the Doubter, coolly, at the same time shrinking back a little. "Likely you will, if you look in the right direction. Keep your dander down till then, and you'll see a good deal better. In the mean time, gentlemen, if you like to listen, I'll tell you what happened when the dyspepsia was bad on me."

Of course, any proposition calculated to restore harmony was heartily approved, and thereupon we were forced to listen to—

CHAPTER XX.

THE DOUBTER'S DYSPEPTIC STORY.

Once, when the dyspepsia was bad on me, I went to bed raything low-spirited, and began to think I was going to die. I thought I couldn't live till morning. My stomach was as hard as a brick-bat, and I was cold all over. The more cover I piled on, the colder I got. The minute I shut my eyes, I was scared to death at the darkness. I felt as if something dreadful was going to

THE DOUBTER.

happen, and didn't know exactly what it was. Sometimes I thought robbers were under the bed, and sometimes I heard strange noises about the house. My heart stopped beating altogether; I felt for my pulse, but couldn't find it in my wrists or any where else. Every bit of blood seemed to have oozed out of me in some mysterious way, and to all intents and purposes my body was dead. There was no dream about it. I could move my limbs the same as ever, and was as wide awake as I am this minute; but there was no sign of life about me except that my mind had power to move the dead flesh; for it was cold and clammy as that of a corpse. Any body else would have given up, and concluded he was a genuine corpse; but you see I was not the sort of man to believe such a thing as that without farther proof. I therefore lay still a while, in hopes I'd get warm by-and-by, and feel better; but I kept growing colder and colder, and at last was so cold that I felt like ice all over. I had the most dreadful and gloomy reflections. Every thing I thought about seemed blue, and dreary, and hopeless; every body unhappy; and the whole future a desert waste, without one ray of light. Despair was upon me; I cared for nothing; it was all the same to me whether I lived or died. I wanted neither help, nor pity, nor love, nor life — all, all was wrapped in despair. The gloom of this state brought on a kind of lethargy; a total unconsciousness of every thing external. My mind only existed and operated, as it were, in perfect darkness. The body was nothing but a type of intense darkness and coldness wrapped around the spirit. In this state I at length heard whisperings in the air, outside of me as

I thought. They drew nearer; the voices were strange and unnatural; I was conscious of a singular sensation, for a time, as if whirled rapidly through space; then I heard the voices say, in low tones, "How cold he is! how miserably cold he is! but we'll soon warm him!" I now became sensible of strong gases in the air, but they produced no farther impression than the mere consciousness of their existence. Wild shrieks and moans, and dreadful hissing sounds arose around me. "Here we are," said the voices; "glad of it, for he's terribly cold." "Put him there in that big furnace; it'll soon warm him," said another voice, in a tone of authority. I was then tossed, as I thought, some distance, and became suddenly still; but the same cold and impenetrable darkness was around my spirit. "There, that fire's out!" said the voice, angrily; "put him in another, and keep him well stirred up." Again there was a movement, and again I was still, but not so still as before, for I was conscious of a jarring sensation. "Out again!" roared the same voice, fiercely. "Out again! you don't keep him well stirred up!" "He's as cold as ice," said the other voices; "we can't do any thing with him." "Try him in the middle furnace!" said the chief voice, sternly; "that'll melt the ice out of him!" Again I was whirled through the gases and deposited in some imperceptible place; but all this time I was growing colder and colder. There was a pause, and then the voices said, "He won't burn, sir; don't you see he's putting the fire out." "Out again, by all the demons!" roared the chief voice, furiously. "Take him away! Carry him back to where you got him. The man's dyspeptic. We can't have such a miserable wretch here! By Pluto! he'd put out every fire we've got in a week. Bear a hand, you rascals! for may I be blessed *if I ain't freezing myself!*" Here the Doubter paused.

"Well, sir, well," said Abraham, ironically, "have you any thing further to say on the same subject? any thing equally reliable? Perhaps you can inform us how you got warm again?"

"Well, that doesn't properly belong to the story," said the Doubter, looking around meaningly upon the company. "I meant that it should end there; but, if you insist upon it, I'll answer your question."

"Of course, sir; the matter requires explanation. It comes to rather an abrupt conclusion."

"The way I got warm, then, was this: I picked up a skull when I was leaving the premises. It was full of hot glue. The fellows that were carrying me got their hands frostbitten and had to let go at last. I fell on an island. The first thing I struck was the top of a mountain. I slid down for three miles without stopping. On the way I broke the skull, and spilled the glue all over me, which made me slip so fast that I was quite warm by the time I got to the bottom."

To this Abraham made no reply. Turning away from the Doubter with ferocity and indignation depicted in every feature, he looked silently around upon the company; his breast heaved convulsively; his hands grasped nervously at the hair upon his goatskin; he deliberately tore it out by the roots; he suppressed a rising smile upon the face of every individual in the party by one more look at the Doubter—one terrible, scathing, foreboding look of vengeance on the morrow; and then said, in a suppressed voice, "Gentlemen, suppose we turn in; it must be twelve o'clock."

CHAPTER XXI.

BAD DREAM CONCERNING THE DOUBTER.

As well as we could judge, Abraham was right in regard to the time; and being all tired, after the story of the dyspeptic man we set about arranging our quarters for the night. I must admit, however, take it all in all, not omitting even the drawbacks to our enjoyment occasioned by the unfortunate state of things between my friend and the Doubter, and the probability of a hostile meeting in the morning, that from the time of leaving home, four months before, I had not spent so pleasant an evening. It was something to look back to with gratification and enjoyment all the rest of the voyage, should we indeed ever be able to resume our voyage.

Pearce now pulled down an additional lot of goatskins from the rafters, which we spread on the ground so as to make a general bed; and having piled some wood on the fire and bolted the door, we stretched ourselves in a circle, with our feet toward the blaze, and made a fair beginning for the night. It was only a beginning, however, so far as I was concerned, for not long after I had closed my eyes and begun to doze, some restless gentleman got up to see if there was any Spaniards trying to unbolt the door; and in stepping over me he contrived to put one foot upon my head, just as I was trying to get from under a big rock that I saw rolling down from the top of a cliff. I was a good deal astonished, upon nervously grasping at it, to find that it was made of leather, and had a human foot in it, and likewise that it had a voice, and asked me, as if very much frightened, "What the deuce was the matter?" This again, upon falling into another doze, brought to mind the footprint

THE FOOTPRINT IN THE SAND.

in the sand, which occasioned me the greatest distress and anxiety. I tried to get away from it, but wherever I went I saw that fatal mark; in the mountains, in the valleys, in the caves, on the rocks, on the trees, in the air, in the surf, in the darkness of the storm, I saw that dreadful footprint; I saw it, through the dim vista of the past, upon the banks of the Ohio, where I had played in boyhood; I saw it again in my first bright glowing dream of the island world, when, with the simplicity of childhood, I prayed that I might be cast upon a desolate island; I saw it in the cream-colored volume—every where—back in childhood, in youth, now again in manhood—from the first to the last, at home, abroad—wherever thought could wander, I saw that strange and wondrous footprint.

In trying to get up the cliff where I could look out for the savages, I fancied the tuft of grass that I had hold of gave way, and I rolled over the precipice into the sea; and this was not altogether an unfounded idea, for I actually had worked myself off the goatskin, and was at that moment paddling about in a sea of mud. Again I fell asleep, and a great many confused visions were impressed upon my mind. I saw the savages down on the beach, going through all their infernal orgies.

THE SAVAGE ORGIES.

They had seized upon my comrades, and were roasting them in flaming fires, and eating the fattest of them with great relish. The flesh of the Doubter, I thought, was so lean and tough that they were unable to eat it; but they stripped it off in long flakes, and hung it round their necks, and danced with it swinging about their bodies, as if they regarded it as the finest ornament in the world. His head was cut off and scalped, and his skull lay upon the ground. I thought Abraham had changed again into Friday, and I called upon him to look at this dreadful scene, and help me to kill these wretched cannibals; but no sooner did he catch sight of the Doubter's skull, than he ran from me toward the spot, and picked it up with a horrible shout of triumph, and sticking his gun into it he held it in the air, and danced all round in a circle laughing like a devil. The Doubter, perceiving this in some strange way (for he was without a head), jumped to his feet, with his fleshless bones, and ran after Abraham, making signs for his skull; but Abraham only laughed the louder and danced the more, thrusting the skull at him as he jumped about, and asking him, in a sneering voice, what he thought of it now? was it a dog's skull yet? would he like to have it fastened on again with glue? how had he contrived to keep out of the fire? were the savages afraid he would put it out? did his present exercise warm him? each of which taunting questions he ended with a wild laugh of derision, and a snatch of his favorite song—

"Tinky ting tang, tinky ting tang,
Oh, poor Robinson Crusoe!"

This, I thought, so incensed the Doubter that he turned away in disgust, and walked off shaking his neck as if it had the head still on; and when he was some distance from Abraham he sat down on the ground and slowly raised his right hand, placing the thumb where his nose would have been had the head still remained in its place, and then his left hand in the same way, fixing the thumb upon the little finger of the other, and thus he waved them to and fro, as if he had no confidence even in his own skull or in any of the circumstances connected with it. While this was going on, the savages continued their infernal dance on the beach. I now raised my gun and began shooting at them, killing them by scores. I could see their dark bodies roll over into the surf, and hear their yells of terror at the report of the gun; and when I rushed down to save my shipmates, all I could see was Abraham sitting upon a rock, pounding the skull into small fragments with a big stone which he held in both hands, and the fleshless body of the Doubter sitting opposite to him, slowly waving the little finger of his left hand at him in the same incredulous and taunting manner as before. And thus ended the dream.

CHAPTER XXII.

THE UNPLEASANT AFFAIR OF HONOR.

When I awoke it was daylight. My mind was still harassed with the bad dream concerning the Doubter. I had the most gloomy forebodings of some impending misfortune either to him or my friend Abraham. Every effort to shake off this unpleasant feeling proved entirely vain; it still clung to me heavily; and, although I was now wide awake, yet it seemed to me there was something prophetic in the dream. Unable to get rid of the impression, I got up, and looked around upon my comrades, who were all sleeping soundly after their rambles of the previous day. Instinctively, as it were, for I was unconscious of any fixed motive, I counted them. There were only nine! A sudden pang shot through me, as if my worst fears were now realized. But how? I thought. Where was the tenth man? What had become of him? Was it Abraham? Was it the Doubter? Who was it? for the light was not strong enough to enable me to distinguish all the faces, partly hidden, as they were, in the goatskins. I looked toward the door; it was unbolted, and slightly ajar. I opened it wide and looked out; there was nothing to be seen in the gray light of the morning but the bushes near the hut, and the dark mountains in the distance. It was time, at all events, to be on the look-out for the ship, so I roused up my comrades, and eagerly noticed each one as he waked. The Doubter was missing! Could it be possible that Abraham's threats had driven him to run away during the night, when all were asleep, and hide himself in the mountains? There seemed to be no other way of accounting for his absence. "Where is he? what's become of him? maybe he's drowned himself!" were the general remarks upon discovering his absence. "Come on! we must look for him! it won't do to leave him ashore!" We hurried down to the boat-landing as fast as we could, thinking he might be there; and on our way saw that the ship was still in the offing. The boat was just as we had left it, but not a soul any where near. We then roused up every body in the Chilian quarter, shouting the name of the missing man in all directions. He was not there! All this time Abraham was in the greatest distress, running about every where, without saying a word, looking under the bushes, peeping into every crevice in the rocks, darting in and out of the Chilian huts, greatly to the astonishment of the occupants, and quite breathless and dispirited when he discovered no trace of our comrade. At last, when we were forced to give up the search and turn toward Pearce's hut, where we had left our host in the act of lighting the fire to cook breakfast, he took me aside, and said, "Look here, Luff, I'm very sorry I had any difficulty with that poor fellow. The fact is, he provoked me to it. However, I have nothing against him now; and I just wanted to tell you that I sha'n't go aboard the ship till I find him. If you like, you can help me to hunt him up, while the others are seeing about breakfast."

"To be sure, Abraham," said I, "we must find him, dead or alive. I'll go with you, of course. But tell me, as we walk along, what it was Pearce said to you last night. How did you get him back when he went out?"

"Oh, never mind that now," replied Abraham, looking, as I thought, rather confused.

"You gave him a dollar, didn't you?" said I; "what was that for?"

"Why, the fact is, Luff, he made those marks himself in some idle hour as he lay basking in the sun up there. He told me that he often spends whole days among the cliffs or sleeping in the caves, while his sheep are grazing in the valleys. You may have noticed that he was rather inclined to burst when he left the hut. The fellow had sense enough not to say any thing before the company. I thought it was worth a dollar to keep the thing quiet."

"It was well worth a dollar, Abraham; but the skull—what about the skull?"

"Oh, the skull? *He* said he picked it up one day outside the cave, and hove it up there, thinking it would do for a lamp some time or other. What excited me so when our shipmate spoke about it was that he should call it a dog's skull."

"And wasn't it?"

"Why, yes; to tell the truth, Luff, it *was* the skull of a wild dog; but you know one doesn't like to be told of such a thing. However, we must look about for the poor fellow, and not leave him ashore."

By this time we had reached an elevation some distance back of the huts. We stopped a while to listen, and then began shouting his name. At first we could hear nothing; but at length there was a sound reached our ears like a distant echo, only rather muffled.

"Halloo!" cried Abraham, as loud as he could.

"Halloo!" was faintly echoed back, after a pause.

"Nothing but an echo," said I.

"It doesn't sound like *my* voice," observed Abraham. "Halloo! where are you?" he shouted again, at the highest pitch of his voice. There was another pause.

"I'm here!" was the smothered reply.

"That's a queer echo," said Abraham; "I'll bet a dollar he's underground somewhere. Halloo! halloo! Where are you?" This time Abraham put his ear to the ground to listen.

"Here, I tell you!" answered the voice, in the same smothered tones. "Down here."

"He's not far off," said Abraham. "Come, let us look about."

We immediately set out in the direction of the voice. The path made a turn round a point of rocks some few hundred yards distant, on the right of which was a steep precipice. On reaching this, we walked on some distance, till we came to a narrow pass, with a high bluff on one side, and a large rock on the edge of the precipice. The path apparently came to an end here; but upon going a little farther, we saw that it formed a kind of step about three feet down, just at the beginning of the narrow pass, between the rock and the bluff, so that in making any farther progress it would be necessary to jump from the top of the step, or, in coming the other way, to jump up. It was necessary for us, at least, to jump some way before long, for upon arriving at the edge we discovered a pit about four feet wide at the mouth, and how deep it was impossible for us to tell at the moment. We thought it must be rather deep, however, from the sepulchral sounds that came out of it. "Here I am," said the voice, "down in the hole, here, if I ain't mistaken, but I wouldn't swear to it; I may be somewhere else: it feels like a hole—that's all I can say about it, except that it's tolerably deep, and smells of goats."

"A goat-trap!" exclaimed Abraham, in undisguised astonishment. "By heavens, Luff, he's caught in a goat-trap!"

"It may be a goat-trap, or it may not. I want you to observe that I neither deny nor affirm the proposition. There's not much room in it, however, except for doubt."

"How in the world are we to get him out?" cried Abraham, whose sympathies were now thoroughly aroused by the misfortune of his opponent. "We must contrive some plan to pull him out. Hold on here, Luff; I'll go and cut a pole."

While Abraham was hunting about among the bushes for a pole of suitable length, I sprang over to the other side of the pit, and, getting down on my hands and knees, looked into it, and perceived that it spread out toward the bottom, so that it was impossible to climb up without assistance.

"This is rather a bad business," said I; "what induced you to go down there?"

"I didn't come down here altogether of my own will," replied the Doubter; "credulity brought me here—too much credulity; taking things without sufficient proof; assuming a ground where no ground existed."

"How was that? I don't quite understand."

"Why, you see, I happened to come along this way about an hour ago, to see if the sun rose in the north, and not dreaming of goat-traps, I took it for granted that I could jump down a step in the path apparently not more than three feet deep. There's where the mistake was. A man has no business placing any dependence upon his eyes without strong collateral evidence from all the rest of his senses. I assumed the ground that there was ground at the bottom of the step. Accordingly, I jumped. There was no ground for the assumption. To be sure I descended three feet, according to my original design; but I descended at least twelve feet more, of which I had no intention whatever. The fact is, there was some rotten brushwood, covered with straw and clay, over the mouth of the pit, which I went through without the least difficulty."

"Are you hurt?" said I, anxiously.

"Well, I was considerably stunned. Likely enough some of my ribs are broken, and several blood-vessels ruptured; but I won't believe any thing more for some time. I've made up my mind to that. I may or may not be hurt, according to future proof."

By this time Abraham came running toward the pit as fast as he could, with a long pole in his hand, which he had cut among the bushes.

"This is the best I could get," said he, nearly breathless with haste, and very much excited; "there were some others, but I didn't think they were strong enough." Without farther delay, he sprang across the pit to the lower side, and thrust the pole down as far as he could reach. It must have struck something, for he immediately drew it back a little, and the voice of the Doubter was heard to exclaim, in a high state of irritation,

"Halloo, there! What are you about? Confound it, sir, I'm not a wild beast, to be stirred up in that way."

"Never mind," said Abraham, "I didn't intend to hurt you. Take hold of the pole. I'll pull you out. Take hold of it quick, and hang on as hard as you can."

"No, sir; it can't be done, sir. I'll not take hold of any thing upon an uncertainty."

"But there's no uncertainty about this," cried Abraham, in a high state of excitement; "it's perfectly safe. Take hold, I tell you."

"Can't be done, sir, can't be done," said the Doubter; "there's not sufficient proof that you'll pull me out if I do take hold. No, sir; I've been deceived once, and I don't mean to be deceived again."

"Now, by heavens, Luff, this is too bad. He doubts my honor. What are we to do?" And Abraham wrung his hands in despair. "Halloo, there, I say—halloo!"

"Well, what do you want?" answered the voice of the Doubter.

"I want to pull you out. Surely you don't think I'll be guilty of any thing so dishonorable as to take advantage of your misfortune?"

"I don't think at all," said the Doubter, gloomily;

"I've given up thinking. You may or may not be an honorable man. At present I have nobody's word for it but your own."

Here I thought it proper to protest that I knew Abraham well; that there was not a more honorable man living. "Besides," I added, "there's no other way for you to get out of the pit."

"Very well, then," said the Doubter; "I'll take hold, but you must take hold too, and see that he doesn't let go. Pull away, gentlemen!"

Abraham and myself accordingly pulled away as hard as we could, and in a few moments the head of our comrade appeared in the light, a short distance below the rim of the pit. I had barely time to notice that his hair was filled with straw and clay, when Abraham, in his eagerness to get him entirely clear of danger, made a sudden pull, which would certainly have accomplished the object had the Doubter come with the upper part of the pole. But such was not the case. On the contrary, both my friend and myself fell flat upon our backs;

THE DOUBTER BACK AGAIN.

and upon jumping up, we discovered that the Doubter had fallen into the pit again, carrying with him the lower end of the pole, which had unfortunately broken off at that critical moment. There he lay in the bottom of the pit, writhing and groaning in the most frightful manner.

"He's killed! he's killed!" cried Abraham, in perfect agony of mind. "Oh, Luff, to think that I killed him at last! It was all my fault. Here, quick! Lower me down! I must help him!"

Before I had time to say a word, Abraham seized hold of my right hand, and, directing me to hold on with all my might, he began to let himself down into the pit. It required the utmost tension of every muscle to bear his weight, but the excitement nerved me. "Let go, now!" said he, as soon as he got as far down as I could lower him without lowering myself, which I narrowly escaped; "let go, Luff!" I did so, and heard a dull, heavy fall, and a groan louder than before.

"What's the matter, Abraham—did you hurt yourself?"

"Not myself," said Abraham, "but I'm afraid I hurt him. I fell on him."

"You did," groaned a voice, faintly, "you fell on me. I'm tolerably certain of that. It was a shabby trick, sir; it wasn't bad enough to throw me down here, without jumping on top of me when I couldn't defend myself!"

"I hope you're not much hurt," said Abraham; "it was all accident—I swear it, on my sacred honor!"

"Honor!" groaned the Doubter, contemptuously; "is it honorable to drop a man into a pit, and knock all the breath out of his body, and then jump on top of him! Honor, indeed! But it was my own fault: I was too ready to take things without proof."

"Now, by all that's human!" cried Abraham, stung to the quick at these unmerited reproaches, "I'll prove to you that I didn't mean it. Get up on my shoulders—here, I'll help you—and climb out. Would any but an honorable man do that?"

"It depends upon his motives," replied the Doubter; "I won't take motives on credit any more. I'm not going to get up on your shoulders, and have you jump from under me about the time I get hold of something above, and leave me to fall down and break my back, or hang there. No, sir, I want no farther assistance. I've made up my mind to spend the remainder of my days here."

"You *sha'n't* stay here!" cried Abraham, exasperated to the last degree by these taunts. "By heavens, sir, you *shall* be assisted!"

Here there was a struggle in the bottom of the pit; the Doubter writhing like an eel all over the ground in his attempts to elude the grasp of Abraham; but soon he was in the powerful arms of my friend, who, holding him up, shouted lustily, "Catch hold of him, Luff! Catch him by the hair or the coat-collar! Hold on to him, while I shove him up!"

The writhing form of the Doubter at the same moment loomed up in the light, and I called upon him to give me his hands; but he resolutely held them down, protesting that he would trust no man for the future; that he'd die before any body should deceive him again. In this extremity, driven almost frantic in my zeal for his safety, I grasped at the collar of his coat, and succeeded, after some difficulty, in getting a firm hold of it. "All right!" I shouted; "push away now, Abraham!" In spite of every exertion on Abraham's part, however, our unfortunate comrade rose no higher, which I can only account for by the depth of the pit. "A little higher, Abraham—just two inches—that's it—all right!" It certainly was all right so far; I had drawn him partly over the edge, and would eventually have drawn him entirely over, had it not caved in, by reason of the united weight of both on it at the same time, and thus the matter was prevented from being all right to any greater extent. The consequence of this disaster was, that we both fell heavily upon Abraham, who, unable to bear our united weight, fell himself under the Doubter, while I, being uppermost, formed a kind of apex to the pyramid.

Our fall was thus broken in some measure; and, although Abraham groaned heavily under our weight, yet, as fortune would have it, nobody was hurt. The Doubter was the first who spoke.

"I told you so!" said he, faintly; "but you *would* try. You would try, in spite of all I could say, and now you see the consequence. It appears to me that there are three men caught in a goat-trap now instead of one; but I'll not insist upon it; there may be only one. My eyes have deceived me already, and likely as not they deceive me now."

"No, they don't," said Abraham, in smothered tones; "I'm quite certain there are two of you on top of me. Get off, if you can, for I can't breathe much longer in this position. You may depend upon it, there are three of us here."

"I shall depend upon nothing for the future," replied the Doubter, gloomily; "I depended upon a pole just now, and was dropped; I put faith in that pole, and both the faith and the pole were broken at the same time, and my back too nearly, if not quite broken."

"But I'm not a pole," groaned Abraham, "you may depend upon that. Get off now, do, for heaven's sake."

"You don't feel like a pole," said the Doubter, "but you may be one, for all I know; there's no telling what you are. However, I'll get off, lest you should break likewise."

I had already relieved Abraham of my weight; and being now entirely free, he got up, and we began to consider how we were to get out of the pit.

As good luck would have it, we heard some voices approaching, which we soon discovered to be a couple of Chilians, to whom the trap belonged, coming thus early in the morning to see if it had caught any goats. When they looked over and saw the earth broken in, they were greatly rejoiced; but no sooner did they perceive that the game consisted of three full-grown men, than they ran away as fast as they could, shouting "*Diabolo!*

Diabolo!" Abraham, who had been studying Spanish during the voyage, understood sufficient of the language to call out "*Americanos! Americanos! no Diabolo! Per amore Deos, viene' qui! Amigos! amigos! no Diabolo!*" This caused them to halt; and upon its being repeated a great many times, they ventured to the edge of the pit, where Abraham gave them every assurance that we were three unfortunate Americans, who had fallen into the trap by accident, and that we were in no way related to the devil. Upon this, they took a coil of rope, which they had for pulling up goats, and making a noose on one end, they let it down. The first man that was fastened on was the Doubter. It required the united efforts of Abraham and myself to get him into the noose; but we eventually had the pleasure of seeing him go up through the hole without farther accident. I then yielded reluctantly to Abraham, who insisted, as a point of honor, that he should be the last man. Being light, I was whirled out in a twinkling; and, finally, through this providential turn of affairs, we were all safely landed outside of the pit. The two Chilians, unable to divine the causes which had led to this singular state of things, looked on as if still half afraid that they had pulled some very bad characters out of the ground, muttering, as we shook the dirt off our clothes, "*Madre de Deos! Santa Maria! Padre bonita!*" I considered this a fitting opportunity, in view of the happy issue of the disaster, to effect a full and complete reconciliation between Abraham and the Doubter, and therefore proposed that they should shake hands on the spot, and forego all future hostilities. My friend immediately held out his hand in the frankest manner; the Doubter hesitated a moment, as if afraid that it might result in his being pulled back again into the pit; but, unable any longer to resist the hearty sincerity of his opponent, he gave his hand, and suffered it to be shaken; and so rejoiced was Abraham in finding every thing was thus happily settled, that he shook on with all his force for at least five minutes, during which the two Chilians, knowing no good reason why a pair of strange gentlemen, just pulled out of a goat-trap, should stand shaking hands with one another, exhibited the utmost surprise and consternation, exclaiming, as before, "*Madre de Deos! Santa Maria! Padre bonita!*"

We contrived to make up the sum of a dollar between us, which we gave to the men, telling them, at the same time, that they need not mention this matter, should they see any of our companions before we left the island. We then started for Pearce's hut, which we soon reached. The rest of the party had finished breakfast, and were waiting for us at the boat-landing. They had left directions with Pearce that we were to follow without delay, with or without the missing man, as the ship had made a signal for us to come aboard. While the Doubter and myself were making a hasty snack, Abraham took a piece of bread and meat, and started off to let our friends know that we had found the missing man, and would soon be down. In a few minutes we concluded our snack, and were about leaving the cabin, when Pearce said he reckoned some of us had left a bundle, which he had found in the corner. The bundle consisted of a handkerchief tied up, with something in it, which I quickly discovered to be the relic we had found in Crusoe's Cave.

"Where did you get that?" said Pearce.

"We dug it up in Crusoe's Cave; it was made by Alexander Selkirk."

"No it wasn't; it was made by me. I lived there a while when I first came on the island, and made it myself. I know the mark. I made it about a year and a half ago."

"But how is that?" said I, greatly astonished; "it looks to be over a century and a half old."

"It wasn't baked enough," said Pearce; "that's the reason it didn't keep well. The name's broke off, but there's part of what I writ on it."

"Impossible!" said I. "Don't you see 'A S........

170–?' What can that be but Alexander Selkirk, 1704, which was just the time he lived here!"

"No, 'taint; Alexander Selkirk never made that 'ere. I made it myself. I put my name on it; but the name's broke off. I writ, '*A Saucepan maid by W. Pearce, 17 Oct.*' That's all. 'Taint no use to me now; you may take it, ef you want to."

I took it without saying another word; tied it up again in the handkerchief, and asked Pearce if he was going down with us to the boat-landing. He said he would be down there presently. So, without farther delay, we set out to join our companions. As we walked rapidly along the path, my shipmate suffered strange sounds to escape from his throat, indicative of his feelings. Suddenly he stopped, as if unable to restrain himself any longer.

"Where are you going?" said he.

"Going aboard, to be sure; come on, they're waiting for us."

"You are, eh? going aboard, eh? Well, any thing to humor the idea. It sounds very like reality, indeed—very."

"And why shouldn't it?" said I.

"Of course, why shouldn't it? Look here, Luff, you're rather a clever sort of fellow."

"Do you think so?" said I, a little embarrassed at so abrupt an opinion in my favor.

"Yes, I do," said the Doubter; "I always did. Will you just have the goodness to look into my mouth (opening it at the same time as wide as he could). Now, just cast your eyes into this cavity."

I did as he desired me, thinking perhaps the poor fellow was suffering from his fall into the goat-pit.

"Well," said I, "there's nothing there, so far as I can see, except a piece of tobacco. Your tongue looks badly."

"It does, eh? No matter about that. This is what I want you to notice: that I have a tolerably big swal-

SWALLOWING AN ISLAND.

lowing apparatus, but I'm not the style of man that's calculated to swallow an entire island. Possibly I might get down a piece of a skull, or an old saucepan, with a grain of salt; but I can't swallow Juan Fernandez, with Robinson Crusoe and Alexander Selkirk—two of the biggest liars that ever existed, on top of it. No, sir, it can't be done."

I thought myself that he was not a person likely to accomplish a feat of that kind, for his throat was not uncommonly large, and his digestive organs appeared to be weakly.

"No, I shouldn't think so," said I. "You don't look like a man that could swallow so much."

"Very well, then; I'm willing to humor the idea. I'll imagine we're going aboard from Juan Fernandez, if you like. But the island *doesn't exist!* No, sir; it reads very well on paper; it's a very romantic place, no doubt—if any body could find it; a very pleasant spot for a small tea-party between a pair of wandering vagabonds; but it doesn't exist any where else but on the maps. Don't you ever try, Luff, to make me believe that any of these things which we imagine to have occurred within the past three days have the slightest foundation in fact."

I was not prepared to go to the full extent of denying the entire existence of the island; but, I must confess, there was a good deal in our experiences of the past three days calculated to inspire doubt; so much, indeed, that I hardly knew what to believe myself. Even now, after the lapse of four years, and the frequent repetition of all these adventures to my friends, which has given something more of reality to the doubtful points, I would hardly be willing to swear to more than the general outline; nor am I quite certain that even the main incidents would stand cross-examination in a Court of Doubters. Such, reader, is the deceptive nature of appearances!

While we were talking, Pearce overtook us with a bundle of goatskins which we had bargained for the night before, and we all went down to the boat-landing together. There we found our shipmates all ready to start. The Anteus was lying-to about eight or ten miles off, outside the harbor; and the sea being rather rough, we thought it best to agree with Pearce for some seats in his boat, and hire a couple of the Chilians to help us at the oars. In this way, having stored all our relics in the bow of the boat except the earthen pot, which we had the misfortune to drop overboard, we set out for the ship, bidding a general good-by to Juan Fernandez and all its romantic vales with three hearty cheers. A few heavy seas broke over us when we got outside the harbor; and we saw the Brooklyn weighing anchor and preparing to stand out to sea, and a small brig that we had met in Rio beating in; but, with the exception of these little incidents, nothing occurred worth mentioning till we arrived alongside the Anteus. The captain and all the passengers received us in silence; not a word was spoken by any body; no sign of rejoicing or recognition whatever took place as we stepped on board. We thought it rather a cool termination to our adventures, and could only account for it by supposing that this was the way people thought to be dead and buried are usually treated when they come unexpectedly to life again after a great deal of grief has been wasted upon them. Nor were we wrong in our conjectures; for in about five minutes our friends on board, including the kind-hearted captain, finding themselves entirely unable to keep up such a state of displeasure, crowded around us in different parts of the ship, and began shaking hands with us privately, and asking us a thousand questions about Juan Fernandez and Robinson Crusoe. We introduced our worthy host as the real Crusoe of the island, and brought both him and the Chilians down into the cabin, where we gave them as much as they could eat, besides honorably acquitting ourselves of our indebtedness by paying our friend Pearce all the ham and bread we had promised him, and loading him with sundry presents of clothing and groceries. The captain then ordered the yards to be braced; the boat swung off as we began to plow our way once more toward the Golden Land, and before noon the island was blue in the distance.

CHAPTER XXIII.

DOCTOR STILLMAN'S JOURNAL.

I HAVE been kindly permitted to select the following from the private journal of Dr. J. D. B. Stillman, of New York, an intelligent fellow-passenger on the Anteus. It will give some idea of the state of feeling on board during our absence.

"*Sunday, May 20th.* Eleven passengers left the vessel yesterday in a small boat, with the intention of going ashore on the island of Juan Fernandez for fruit and fresh provisions. At first they made but little progress ahead of the ship, but the wind soon fell away entirely, and about noon the boat could not be seen from the mast-head. Another party of eight passengers prepared to start about two o'clock this morning. The captain, however, was so uneasy at the absence of the other boat, that he refused liberty. Lights were kept burning in the rigging during the night. Toward morning a breeze sprung up. Short sail was carried for fear the boat should attempt to reach us and miss her way. At sunrise it was again calm. The islands loomed higher, but nothing could be distinguished. At 11 A.M. a stiff breeze sprang up from the direction of Masatierra, and the day was spent in beating to windward, and straining our eyes in the hope of discerning some traces of our lost comrades. The wind continued to freshen all day. At 8 P.M. the sea was quite rough. No light could be seen on the shore. The captain, who is well acquainted with the island, says if they attempted to land on the south side they would be inevitably swamped, and some or all lost, as the shore is rock-bound, and the only safe landing is on the north side, fifteen miles farther on. The probability is that they were too much exhausted to attempt landing, and night would have fallen before they could have reached the land at any rate. I am confident in the opinion that they are on the north side of the island, and that they lay all last night on their oars, and landed this morning, too much exhausted to attempt returning the same day. I have great confidence in some of the company; but to-night gloom is general, and a fearful presentiment seems to rest upon the minds of all that we shall soon have to record a melancholy casualty.

"*Monday*, 21*st*. The wind this morning is blowing very fresh. We have been all day beating nearer the island. Objects are quite distinct on the south shore. It is very high and nearly barren. Indeed, so steep are the lofty mountain sides that there does not appear to be soil enough adhering to the rocks to support a spire of grass, except near the summits, which are over a thousand feet in height where they rise near the water; and every where, so far as we can see, the shore is rock-bound, upon which the surf beats fearfully. They could not be so wild as to attempt landing on this side. To-night the wind blows a gale, and we shall be compelled to await a change before we attempt the windward side. Hopes are getting faint. The distress of those who are most interested in the parties is great. Some of our best men were of the company. In fact, it is a question which has absorbed all others, What has become of the boat? To-night I have rather congratulated myself that I did not go. To add to our perplexity, the air is becoming thick, and rain is coming on. The clouds hang heavy and dark over the mountains. At nightfall the wind suddenly changes to S.W. The ship is put about, and run for the north side of the island.

"*May 22d.* While I was writing last night, a loud shout called us all in great haste on deck. A light had been discovered on the shore, and hearty cheers expressed the deep anxiety of all, now in a great measure relieved. There was no doubt that they had reached the shore, and that some of the number were surviving. I felt assured that all was right. Signals were set from the rigging, and the vessel lay to during the night. At dawn of day we were twenty miles distant from the island. Made all sail and stood in for the harbor. As we neared the shore, discovered a large ship at anchor, and a brig rounding the western point. Soon after, we distinguished the tiny sail of our lost boat making for the ship. The captain, in order to show a proper resentment for the disobedience of orders, directed that no demonstrations of joy should be made; and, as they came alongside, they were received in silence."

The shades of evening were gathering upon the horizon. A murmur of life arose from the decks, but it fell unheeded upon my ear. For now, and for many days and nights in our dreary voyage, there was no life for me but in the past. I felt that my happiest hours were there.

Once more I turned to look upon the dim island that

was fading away in the south. A steady breeze wafted us onward; the sun's last rays yet lingered in the sky; twilight hung upon the ocean, and its gentle spirit

"Rendered birth
To dim enchantments—melting heaven to earth—
Leaving on craggy hills and running streams
A softness like the atmosphere of dreams."

DREAMS AND REALITIES.

And was this the last of the island-world? was it to be in future years a mere dream of the past? was I never more to behold its wild grottoes and green valleys? was all the romance of life to fade away with it in the twilight? was it, like the cream-colored volume, to reveal enchantments that henceforth could dwell only in the memory?

Fresh, and fair, and wondrous it was in its romantic beauty when the mists were scattered away, and I beheld it for the first time in the glowing light of morning, with the white sea-foam sparkling on its shores, and the birds singing in its groves. How rich the air was with sweet odors; how varied and changing the colors upon the hill-sides; how softly steeped in shadows were its glens and woodland slopes—what a world of romance was there!

I had pressed its sod with my feet; reveled in its streams; lived again my early life in its pleasant valleys; passed some happy hours there with friends from whom I soon must part; and now, what was it? A dim cloud on the horizon, sinking in the sea, fading away in the shadows of night.

I looked again; faintly and more faintly still its mountains loomed above the deep. Weary with gazing, I closed my eyes, and for a moment I saw it again; but it was only in fancy. I looked—and it had passed away! Was it forever?

> "And now the light of many stars
> Quivered in tremulous softness on the air."

Yet not forever is it lost to me; for often in the busy world I pause and think of that dream-land in the far-off seas, and it rises before me as I saw it in the morning sun, all rich and strange in its beauty; and again I wander through its romantic vales, and again it brings back pleasant memories of the cream-colored volume; and as I look once more, startled from my reverie by the hum of life, it fades away as it faded then in the shadows of night, but not forever. Though I never more may behold it with mortal eyes, yet I see it where distance can not dim the sight: it hath not passed away forever.

PEAK OF YONKA.

CHAPTER XXIV.

CONFIDENTIAL CHAT WITH THE READER.

Now that we have finished our ramble together, and formed something of a speaking acquaintance, I hope, my dear reader, that you will not take it amiss if I hold you a moment by the button, and say a word in confidence. It has been so long the custom of adventurers to speak now and then about themselves, that I assume the privilege without farther apology. If I have been so fortunate as to inspire you with a friendly interest in my behalf during our pleasant wanderings in the footsteps of Robinson Crusoe, I am sure you will be glad to learn that it has always been my greatest ambition to prove myself a worthy disciple of that distinguished adventurer. In this view I have, as you may have noticed, adhered to simple facts, and carefully avoided every thing that might be regarded in the light of fiction, though the temptation to indulge in occasional touches of romance was very difficult to resist. Indeed, so thoroughly have I striven to become imbued with the true spirit of Crusoeism, that much which I thought at first a little doubtful myself, now seems quite authentic; and I think, upon the whole, you may rely upon the truthfulness of my narrative. That I was near being lost in an open boat, with ten others, in trying to get ashore on the island of Juan Fernandez, I conscientiously believe; that we did get ashore, and sleep in caves and straw huts, and climb wonderful mountains, and explore enchanting valleys, I will insist upon to the latest hour of my life; that I have endeavored faithfully to describe the island as it appeared to me, and to give a true and reliable account of its psesent condition, climate, topography, and scenery, I affirm on the honor and veracity of a traveler; that in every essential particular it has been my aim to present a faithful picture of life in that remote little world, I will swear to on the best edition of Robinson Crusoe: more than that it would be unreasonable to expect. If, how-

SCENERY OF JUAN FERNANDEZ.

ever, after this candid avowal, you still insist upon having a distinct and emphatic declaration in regard to any doubtful point, all I can say is, that, like the man who made a statement concerning the height of a certain horse, I am ready at all hazards to stick to whatever I said. If I spoke of a mountain as three thousand miles high instead of three thousand feet, why, in the name of peace, let it be three thousand miles; if I killed any sav-

ages, I am sorry for it, but they must remain dead—it is impossible to bring them to life now; if I put some of my own ideas into the heads of others, it must have been because I thought them better adapted to the subject than what those heads contained already, and I hold myself responsible for them; if at any time I imagined myself to be the original and genuine Crusoe, with a man in my service called Friday, I still adhere to it that no Crusoe more certain than he was himself ever existed upon that island; if, in short, there is any one point upon which I have hazarded the reputation of a veracious chronicler of actual events, or a faithful delineator of strange scenes in nature, I hereby declare that I shall most cheerfully return to Juan Fernandez in an open boat with any ten readers who desire to test the matter by ocular demonstration, and thus convince the most skeptical that I have not made a single unfounded assertion.

And now, in the hope that we may meet again, I wish

to leave you a trifling souvenir by which to bear me in mind.

One of the sailors on board the Anteus was kind enough to make me a suit of clothes out of the goatskins that I bought of Pearce. He made them according to a pattern of my own, which I intend some day or other to introduce in the fashionable circles. I stowed them carefully away in my berth, but the rats took such a fancy to them that, by the time I reached California, there was nothing left but the tail of one goat upon which to hang a portrait; and I regret to say the accompanying sketch, taken from memory, affords but an imperfect conception of the suit as I originally appeared in it. I trust the apparent egotism of smuggling my likeness into print in a suit of goatskins, on the pretext of exhibiting the suit itself, will be excused by the absolute necessity of filling it up with something. At the same time, I must be permitted to observe that the stiffness is in the material, and not in the person of the author.

THE AUTHOR À LA ROBINSON CRUSOE.

CHAPTER XXV.

EARLY VOYAGES TO JUAN FERNANDEZ.

THE group known as Juan Fernandez consists of two chief and several smaller islands, situated in the Pacific Ocean, about four hundred miles from the coast of Chili, in latitude 33° 40′ south, longitude 70° west. These islands were discovered in 1563 by Juan Fernando, a Spanish navigator, whose name they bear. The largest—lying nearest to the main land—is that which is commonly known by the name of the discoverer; it is also called Masatierra. The length of this island is about twelve miles, the breadth six or seven. Ninety miles west is the island of Masafuero, so named to distinguish it from Masatierra. Both are composed of lofty mountains; the harbors are small and unsafe, and the shores, for the most part, are rock-bound. The northern aspect, facing toward the equator, is slightly wooded, and the valleys are fertile; but the southern side, toward Cape Horn, is entirely barren. There are two or three large rocks included in the group, the chief of which, lying at the southern extremity of Masatierra, is called Goat Island, from the great number of goats found there.

According to the early navigators, it would appear that these islands must have been visited by the Indians of South America long before their discovery by Juan Fernando, but it was probably only for the purpose of fishing and catching seals.

The first attempt to form a regular settlement was made by Fernando himself, who, elated by his discovery, and the prospect of colonizing the island, endeavored to obtain a patent from the government at Lima. Failing to receive encouragement from the government, he resolved upon forming a settlement himself; and he visited the island soon after, taking with him some families, with whom he resided there a short time. A few goats, which they carried with them from Lima, speedily stocked the island; and this is probably the origin of these animals in Juan Fernandez, as no mention is made of their having existed there before. Eventually the colony was broken up by the superior inducements held out to settlers in Chili, which at this time fell under the dominion of the Spaniards; and the Spanish authorities of Lima still refusing to grant a patent to Fernando, he was forced to abandon all hope of forming another and more permanent settlement.

For many years subsequently this group was the resort of pirates and buccaneers, who found it convenient, in their cruising in the South Pacific, to touch there for wood and water.

Captain Tasman, a Dutch navigator, sailed from Batavia in 1642, and visited Juan Fernandez in 1643. A translation of his narrative, in Pinkerton's Collection, contains an entertaining account of the island at that period. He dwells enthusiastically upon the advantages of its position, the salubrity of the climate, the fertility of the soil, and strongly urges upon the Dutch East India Company the policy of forming a settlement there, as a dépôt for their commerce in the Pacific.

Alonzo de Ovalle, a native of Chili, gives, in his Historical Relation of the Kingdom of Chili, printed at Rome in 1649, a very entertaining account of what he says he "found writ about these islands, in Theodore and John de Bry, in their relation of the voyage of John Scutten."

Ringrose, in his account of the voyages of Captain Sharpe and other buccaneers, mentions that a vessel was cast away here, from which only one man out of the whole ship's company escaped; and that this man lived five years alone upon this island, before he had any opportunity of getting away in another vessel.

Captain Watlin was chased from Juan Fernandez in 1681 by three Spanish ships. He left on the island a

Musquito Indian, who was out hunting for goats when the alarm was given, and was unable to reach the shore before the ship got under way and put to sea. This Indian, according to Dampier, whose narrative I quote, "had with him his gun and a knife, with a small horn of powder, and a few shot, which being spent, he contrived a way, by notching his knife, to saw the barrel of his gun into small pieces, wherewith he made harpoons, lances, hooks, and a long knife, heating the pieces first in the fire, which he struck with his gun-flint, and a piece of the barrel of his gun, which he hardened, having learned to do that among the English." With such rude instruments as he made in that manner, he procured an abundant supply of provisions, chiefly goats and fish. In 1684, three years after, when Dampier again visited the island, they put out a canoe from the vessel, and went ashore to look for the Musquito man. When they saw him, "he had no clothes left, having worn out those he brought from Watlin's ship, but only a skin about his waist." The scene that ensued is quaintly and touchingly described in the simple language of the narrative. "He saw our ship the day before we came to an anchor," says Dampier, "and did believe we were English, and therefore killed two goats in the morning before we came to an anchor, and dressed them with cabbage, to treat us when we came ashore. He came then to the sea-side to congratulate our safe arrival. And when we landed, a Musquito Indian, named Robin, first leaped ashore, and, running to his brother Musquito man, threw himself flat on his face at his feet, who, helping him up and embracing him, fell flat on his face on the ground at Robin's feet, and was by him taken up also. We stood with pleasure," continues the famous buccaneer, "to behold the surprise, and tenderness, and solemnity of this interview, which was exceedingly affectionate on both sides; and when their ceremonies of civility were over, we also that stood gazing at them drew near, each of us embracing him we had found here, who was overjoyed to see so many of his old friends, come hither, as he thought, purposely to fetch him."

Five Englishmen were left on the island at another time by Captain Davis. After the vessel had sailed, they were attacked by a large body of Spaniards, who landed in one of the bays; but, in consequence of the facilities for defense afforded by the cliffs, they were enabled successfully to maintain their position, although one of the party deserted and joined the Spaniards. They were afterward taken away by Captain Strong, of London.

Captain Woodes Rodgers, commander of the Duke and Duchess, privateers belonging to Bristol, visited Juan Fernandez in February, 1709. The original, and perhaps the most authentic account of the adventures of Alexander Selkirk is contained in a very curious and entertaining narrative of the voyage, written by Captain Rodgers himself, from which it appears that when the ships came near the land, a light was discovered, which it was thought must be on board of a ship at anchor. Two French vessels had been cruising in search of Captain Rodgers's vessel, and these vessels they supposed to be lying in wait for them close to the shore. The boats which had started for the shore returned, and preparations were made for action. On the following day, seeing no vessel there, they went ashore, where they found a man clothed in goatskins, looking, as the narrative says, "wilder than the first owners of them." He had been on the island four years and four months. His name was Alexander Selkirk, a Scotchman, who had been master of the *Cinque Ports.* Having quarreled with Captain Stradling, under whose command he sailed, he was left ashore at his own request, preferring solitude on an unknown island to the life he led on board this vessel. Before the boat that put him ashore left the beach, he repented of his resolution, and begged to be taken back again; but his companions cruelly mocked him, and left him to his fate. It was he that made the fire which had attracted the attention of the two privateers. They took him on board, and, being a good officer, well recommended by Captain Dampier, he was appointed mate on board Captain Rodgers's vessel, and taken to England. The account of his adventures during his long residence on the island is supposed to have formed the foundation of Robinson Crusoe, the most popular romance ever published in any language. A brief but very curious and graphic narrative of his adventures was published in London, soon after his arrival in England, under the quaint title of "Providence displayed; or a very surprising Account of one *Mr. Alexander Selkirk*, Master of a Merchant Man called The *Cinque Ports;* who, dreaming that the Ship would soon after be lost, he desired to be left on a desolate Island in the South Seas, where he lived Four Years and Four Months without seeing the Face of Man, the ship being afterward cast away as he dreamed. As also, How he came afterward to be miraculously preserved and redeemed from that fatal Place by two *Bristol* Privateers, called the *Duke* and *Duchess*, that took the rich *Acapulco* Ship, worth one hundred Ton of Gold, and brought it to England. To which is added, An Account of his Birth and Education. His description of the Island where he was cast; how he subsisted; the several strange things he saw; and how he used to spend his Time. With some pious Ejaculations that he used during his melancholy Residence there. Written by his own Hand, and attested by most of the eminent Merchants upon the *Royal Exchange.*" *Quarto*, containing twelve pages.

Lord Anson visited this island in 1741 for the purpose of recruiting his ships, after a succession of melancholy disasters in their passage round Cape Horn. An accurate topographical survey, and a full and most reliable description of Juan Fernandez, may be found in the narrative of that expedition, compiled from Lord Anson's papers, and other materials, by Richard Walter, chaplain of the Centurion. The style of this delightful narrative is admirable for its simplicity; and the information with which it abounds in regard to the topography, climate, and productions of the island, is perhaps the most authentic of the time.

In 1743 Ulloa visited this group. He gives, among many interesting facts, a curious relation of the origin of the dogs which abound there. "We saw many dogs," he says, "of different species, particularly of the greyhound kind; and also a great number of goats, which it is very difficult to come at, artfully keeping themselves among those crags and precipices, where no other animal but themselves can live. The dogs owe their origin to a colony sent thither, not many years ago, by the President of Chili and the Viceroy of Peru, in order totally to exterminate the goats, that any pirates or ships of the enemy might not here be furnished with provisions. But this scheme has proved ineffectual, the dogs being incapable of pursuing them among the fastnesses where they live, these animals leaping from one rock to another with surprising agility."

Don George Juan touched at Juan Fernandez in 1744, and made several observations of its latitude.

Don Joseph Pizarro gives, in his narrative of his voyages, an account of a visit a few years later.

In 1750 the Spanish government founded a settlement on the principal island, and built a fort for the protection of the harbor. In the following year both the fort and the town were destroyed by a violent earthquake. They were afterward rebuilt farther from the shore, and were in good order and inhabited in 1767, when Carteret visited the island. Soon after the settlement was broken up, and the town and the fortifications were abandoned.

The Chilian government established a penal colony on the same spot in 1819, which, according to some authorities, was discontinued, after repeated efforts to maintain it, on account of its expense; according to others, in consequence of a terrible earthquake, by which the houses and fortifications were destroyed.

CHILIAN. CHILIENNE.

When Lord Cochrane visited the island in 1823, as it appears from a synopsis of Howel's Life of Selkirk, there were but four men stationed on it, apparently in charge of some cattle. A lady who accompanied Lord Cochrane gives the following description of its condition and appearance at that time: "The island is the most picturesque I ever saw, being composed of high perpendicular rocks, wooded nearly to the top, with beautiful valleys, exceedingly fertile, and watered by copious streams, which occasionally form small marshes. The little valley where the town is, or rather was, is exceedingly beautiful. It is full of fruit-trees and flowers, and sweet herbs, now grown wild; near the shore it is covered with radish and sea-side oats. A small fort was situated on the sea-shore, of which there is nothing now visible but the ditches and part of one wall. Another, of considerable size for the place, is on a high and commanding spot. It contained barracks for soldiers, which, as well as the greater part of the fort, are ruined; but the flag-staff, front wall, and a turret are still standing; and at the foot of the flag-staff lies a very handsome brass gun, cast in Spain A. D. 1614. A few houses and cottages are still in a tolerable condition, though most of the doors, windows, and roofs have been taken away, or used as fuel by whalers and other ships touching here. In the valleys we found numbers of European shrubs and herbs—'where once the garden smiled.' And in the half-ruined hedges, which denote the boundaries of former fields, we found apple, pear, and quince trees, with cherries almost ripe. The ascent is steep and rapid from the beach, even in the valleys, and the long grass was dry and slippery, so that it rendered the walk rather fatiguing; and we were glad to sit down under a large quince-tree, on a carpet of balm bordered with roses, now neglected, and feast our eyes on the lovely view before us. Lord Anson has not exaggerated the beauty of the place or the delights of the climate. We were rather early for its fruits, but even at this time we have gathered delicious figs, cherries, and pears, that a few days of sun would have perfected. The landing-place is also the watering-place. There a little jetty is thrown out, formed of the beach-pebbles, making a little harbor for boats, which lie there close to the fresh water, which comes conducted by a pipe, so that, with a hose, the casks may be filled without landing with the most delicious water. Along the beach some old guns are sunk, to serve as moorings for vessels, which are the safer the nearer in shore they lie, as violent gusts of wind often blow from the mountain for a few minutes. The height of the island is about three thousand feet."

"With all its beauties and resources," adds the biographer of Selkirk, "the island seemed destined never to retain those who settled on it; whether from its isolated position, at so great a distance from the continent, or from some other cause, is uncertain. Not long after Lord Cochrane's visit, however, it received an accession of inhabitants, some of them English, who settled in it under the protection of the Chilian government."

These islands (Masafuero and Masatierra) have been convulsed by several of those destructive earthquakes which prevail to such an alarming extent on the western coast of South America. In 1751 and 1835 the destruction was unusually great. The earthquake of 1835 was attended by some remarkable phenomena. An eruption burst from the sea, about a mile from the land, where the water was from fifty to eighty fathoms deep. Smoke and water were ejected during the day, and flames were seen at night.

Mr. Richard H. Dana, Jun., who visited Juan Fernandez in November, 1835, on his voyage to California, gives, in his admirable narrative (Two Years before the Mast), the following graphic account of its condition at that period: "I was called on deck to stand my watch at about three in the morning, and I shall never forget the peculiar sensation which I experienced on finding myself once more surrounded by land, feeling the night-breeze coming from off shore, and hearing the frogs and crickets. The mountains seemed almost to hang over us, and, apparently from the very heart of them, there came out, at regular intervals, a loud echoing sound, which affected me as hardly human. We saw no lights, and could hardly account for the sound, until the mate, who had been there before, told us that it was the 'Alerta' of the Spanish soldiers, who were stationed over some convicts, confined in caves nearly half way up the mountain. At the expiration of my watch I went below, feeling not a little anxious for the day, that I might see more nearly, and perhaps tread upon, this romantic, I may almost say classic island. When all hands were called it was nearly sunrise, and between that time and breakfast, although quite busy on board in getting up water-casks, etc., I had a good view of the objects about me. The harbor was nearly land-locked, and at the head of it was a landing-place protected by a small breakwater of stones, upon which two large boats were hauled up, with a sentry standing over them. Near this was a variety of huts or cottages, nearly a hundred in number, the best of them built of mud and whitewashed, but the greater part only Robinson Crusoe-like—of posts nd branches of trees. The governor's house, as it is called, was the most conspicuous, being large, with grated windows, plastered walls, and roof of red tiles, yet, like all the rest, of only one story. Near it was a small chapel, distinguished by a cross; and a long, low, brown-looking building, surrounded by something like a palisade, from which an old and dingy-looking Chilian flag was flying. This, of course, was distinguished by the title of *Presidio.* A sentinel was stationed at the chapel, another at the governor's house, and a few soldiers, armed with bayonets, looking rather ragged, with shoes out at the toes, were strolling about among the houses, or waiting at the landing-place for our boat to come ashore."

Not long after Mr. Dana's visit this settlement was entirely broken up. The houses and fortifications were destroyed by an earthquake, and the penal establishment was discontinued.

From time to time, up to the present date, there have been straggling settlers on this island, but there has been no attempt since 1835 to colonize it permanently until recently. It has been occasionally visited by vessels of different nations for supplies of wood and water, and such vegetable productions as the valleys afford. American whalers have found it a very convenient stopping-place in their cruisings on the coast of Chili and Peru; but of late years, the whales becoming scarce in these seas, they are forced to push their voyages into more remote regions. Many still touch there, however, on their way to and from the northern coast.

At the time of the writer's visit to Juan Fernandez (May, 1849), the gold excitement had but recently broken out, and vessels bound to California had just commenced making it a place of resort for refreshments in their outward voyages. Since that period, it is stated in the newspapers that an enterprising American has taken the island on lease from the Chilian government, and established a settlement upon it of a hundred and fifty Tahitians, with the design of cultivating the earth, and furnishing vessels touching there with supplies of fruit and vegetables.

CHAPTER XXVI.

ALEXANDER SELKIRK AND ROBINSON CRUSOE.

It is stated in Howel's life of Selkirk that the singular history of this man (Alexander Selkirk) was soon made known to the public, and immediately after his arrival in London he became an object of curiosity, not only to the people at large, but to those elevated by rank and learning. Sir Richard Steele, some time after, devoted to him an article in the paper entitled "The Englishman," in which he tells the reader that, as Selkirk is a man of good sense, it is a matter of great curiosity to hear him give an account of the different revolutions of his mind during the term of his solitude. "When I first saw him," continues this writer, "I thought, if I had not been let into his character and story, I could have discovered that he had been much separated from company, *from his aspect and gesture;* there was a strong but cheerful seriousness in his look, and a certain disregard of the ordinary things around him, as if he had been sunk in thought. In the course of a few months," as it appears by the same writer, "familiar converse with the town had *taken off the loneliness of his aspect, and quite altered the expression of his face.*"

"De Foo's romance of Robinson Crusoe was not published till the year 1719, when the original facts on which it was founded must have been nearly forgotten. There is no record of any interview having taken place between Selkirk and De Foe, so that it can not be decided whether De Foe learned our hero's story from his own mouth, or from such narratives as those published by Steele and others."

On this point a biographer of De Foe remarks: "Astonishing as was the success of De Foe's romance, it did not deter the curious from attempting to disparage it. The materials, it was said, were either furnished by or surreptitiously obtained from Alexander Selkirk, a mariner who had resided for four years on the desert island of Juan Fernandez, and returned to England in 1711. Very probably his story, which then excited considerable interest and attention, did suggest to De Foe the idea of writing his romance; but all the details and incidents are entirely his own. Most certainly De Foe had obtained no papers or written documents from Selkirk, as the latter had none to communicate."

Robinson Crusoe, however, can not be considered altogether a work of fiction. Without adhering strictly to the actual adventures of Selkirk, or of the Musquito Indian who preceded him, it gives, in the descriptions of scenery, the mode of providing food, the rude expedients resorted to for shelter against the weather, and all the trials and consolations of solitude, a faithfully-drawn picture from these narratives, and a most truthful and charming delineation of solitary life, with such reflections as the subject naturally suggested. De Foe was the great medium through which the spirit of the whole was fused; it required the splendor of his genius to preserve from oblivion the lessons therein taught—of the advantages of temperance, fortitude, and, above all, an implicit reliance in the wisdom and mercy of the Creator. He presents them in a most fascinating garb, with all the originality of a master-mind; and it detracts nothing from his credit to say that the pictures are drawn strictly from nature.

As Captain Rodgers well observes in his simple narrative of the adventures of Selkirk, "One may see by this that solitude and retirement from the world is not such an insufferable state of life as most men imagine, especially when people are fairly called or thrown into it unavoidably, as this man was; who, in all probability, must otherwise have perished in the seas, the ship which left him being cast away not long after, and few of the company escaped. We may perceive by this story that necessity is the mother of invention, since he found means to supply his wants in a very natural manner, so as to maintain his life, though not so conveniently, yet as effectually as we are able to do with all our arts and society. It may likewise instruct us how much a plain and temperate way of living conduces to the health of the body and the vigor of the mind, both of which we are apt to destroy by excess and plenty, especially of strong liquor, and the variety as well as the nature of our meat and drink; for this man, when he came to our ordinary method of diet and life, though he was sober enough, lost much of his strength and agility."

De Foe does not, as may be seen by reference to the fourth section of "Robinson Crusoe," lay the scene of his narrative in Juan Fernandez. Robinson starts from the Brazils, where he has been living as a planter, on a voyage to the coast of Guinea. Driven to the northward along the coast of South America by heavy gales, the captain of the vessel found himself "upon the coast of Guinea, or the north part of Brazil, beyond the River Amazon, toward that of the River Oronoco, commonly called the Great River; and began to consult with me," says Robinson, "what course he should take, for the ship was leaky and very much disabled, and he was for going directly back to the coast of Brazil. I was positively against that; and looking over the charts of the sea-coast of America with him, we concluded there was no inhabited country for us to have recourse to till we came within the circle of the Caribbee Islands, and therefore resolved to stand away for Barbadoes; which, by keeping off to sea, to avoid the indraught of the Bay or Gulf of Mexico, we might easily perform, as we hoped, in about fifteen days' sail; whereas we could not possibly make our voyage to the coast of Africa without some assistance both to our ship and ourselves.

"With this design we changed our course, and steered away N.W. by W. in order to reach some of our English islands, where I hoped for relief; but our voyage was otherwise determined; for, being in the latitude of 12° 18′, a second storm came upon us, which carried us away with the same impetuosity westward, and drove us so out of the very way of all human commerce, that, had our lives been saved as to the sea, we were rather in danger of being devoured by savages than ever returning to our own country.

"In this distress, the wind still blowing very hard, one of our men early in the morning cried out Land! and we had no sooner run out of the cabin to look out, in hopes of seeing whereabouts in the world we were, but the ship struck upon a sand, and in a moment her motion being so stopped, the sea broke over her in such a manner that we expected we should all have perished immediately; and we were immediately driven into our close quarters to shelter us from the very foam and spray of the sea."

It will be seen from the above that Robinson Crusoe was not wrecked on the island of Juan Fernandez. In all probability he never saw that island. I regret the fact as much as any body can regret it, because I always thought so till I referred more particularly to his history; but a due regard for truth compels me to give the facts as I find them.

"The History of Robinson Crusoe," says the biographer of De Foe, already quoted, "was first published in the year 1719, and its popularity may be said to have been established immediately, since four editions were called for in about as many months, a circumstance at that time almost unprecedented in the annals of literature. It rarely happens that an author's expectations are surpassed by the success of his work, however astonishing it may seem to others; yet perhaps even De Foe himself did not venture to look forward to such a welcome on the part of the public, after the repulses he had experienced on the part of the booksellers; for, incredible as it now appears, the manuscript of the work had been offered to, and rejected by, every one in the trade.

"The author of Robinson Crusoe would be entitled to a prominent place in the history of our literature even had he never given to the world that truly admirable production; and yet we may reasonably question whether the name of De Foe would not long ago have sunk into oblivion, or at least have been known, like those of most of his contemporaries, only to the curious student, were it not attached to a work whose popularity has been rarely equaled—never, perhaps, excelled. Even as it is, the reputation due to the writer has been nearly altogether absorbed in that of his hero, and in the all-engrossing interest of his adventures: thousands who have read Robinson Crusoe with delight, and derived from it a satisfaction in no wise diminished by repeated perusal, have never bestowed a thought on its author, or, indeed, regarded it in the light of a literary performance. While its fascination has been universally felt, the genius that conceived it, the talent that perfected it, have been generally overlooked, merely because it is so full of nature and reality as to exhibit no invention or exertion on the part of the author, inasmuch as he appears simply to have recorded what actually happened, and consequently only to have committed to paper plain matter of fact, without study or embellishment. We wonder at and are struck with admiration by the powers of Shakspeare or Cervantes; with regard to De Foe we experience no similar feeling: it is not the skill of the artist that enchants us, but the perfect naturalness of the picture, which is such that we mistake it for a mirror; so that every reader persuades himself that he could write as well, perhaps better, were he but furnished with the materials for an equally interesting narrative."

A DANGEROUS JOURNEY.

CHAPTER I.

THE CANNIBAL.

In the summer of 1849 I had occasion to visit San Luis Obispo, a small town about two hundred and fifty miles south of San Francisco. At that time no steamers touched at the Embarcadera, and but little dependence could be placed upon the small sailing craft that occasionally visited that isolated part of the coast. The trail through the Salinas and Santa Marguerita valleys was considered the only reliable route, though even that was not altogether as safe as could be desired. A portion of the country lying between the Old Mission of Soledad and San Miguel was infested by roving bands of Sonoranians and lawless native Californians. Several drovers, who had started from San Francisco by this route to purchase cattle on the southern ranches, had never reached their destination. It was generally believed that they had been murdered on the way. Indeed, in two instances, this fact was established by the discovery of the mutilated remains of the murdered men. No clew could be obtained to the perpetrators of the deed, nor do I know that any legal measures were taken to find them. At that period the only laws existing were those administered by the alcaldes, under the Mexican system, which had been temporarily adopted in connection with the provisional government established by General Riley. The people generally were too deeply interested in the development of the gold regions to give themselves much concern about the condition of other parts of the country, and the chances of bringing criminals to punishment in the southern districts were very remote.

MIRAGE IN THE SALINAS VALLEY.

My business was connected with the revenue service. A vessel laden with foreign goods had been wrecked on the coast within a short distance of San Luis. It was necessary that immediate official inquiry should be made into the circumstances, with a view of securing payment of duties upon the cargo. I was also charged with a commission to establish a line of post-offices on the land-route to Los Angeles, and enter into contracts for the carrying of the mails.

By the advice of some friends in San Francisco, I purchased a fine-looking mule recently from the Colorado.

The owner, a Texan gentleman, assured me that he had never mounted a better animal; and, so far as I was capable of judging, the recommendation seemed to be justly merited. I willingly paid him his price—three hundred dollars. Next day, having provided myself with a good pair of blankets, a few pounds of coffee, sugar, and hard bread, and a hunting-knife and tin cup, I bade adieu to my friends and set out on my journey. A tedious voyage of six months around Cape Horn had given me a peculiar relish for shore-life. There was something very pleasant in the novelty of the scenery and the inspiring freshness of the air. The rush of emigrants from all parts of the world; the amusing scenes along the road; the free, social, and hopeful spirit which prevailed among all classes; the clear, bright sky, and wonderful richness of coloring that characterized the atmosphere, all contributed to produce the most agreeable sensations. It was a long and rather hazardous journey I had undertaken, and it would doubtless be very lonesome after passing San Jose; but the idea of depending solely on my own resources, and becoming, in some sort, an adventurer in an almost unknown country, had something in it irresistibly captivating to one of my roving disposition. I had traveled through Texas under nearly similar circumstances, and enjoyed many pleasant recollections of the trip. There is a charm about this wild sort of life, the entire freedom from restraint, the luxury of fresh air, the camp under the trees, with a bright fire and a canopy of stars overhead, that, once experienced, can never be forgotten.

Nothing of importance occurred till the evening of the fourth day. I met crowds of travelers all along the road, singing and shouting in sheer exuberance of spirit; and not unfrequently had some very pleasant and congenial company, bound either to the mines or in search of vacant government land for the location of claims. The road through the valleys of Santa Clara and San Jose was perfectly enchanting, winding through oak groves, and fields of wild oats and flowers; and nothing could exceed the balminess of the air. Indeed, the whole country seemed to me more like a succession of beautiful parks, in which each turn of the road might bring in view some elegant mansion, with sweeping lawns in front, and graceful ladies mounted on palfreys, than a rude and uncivilized part of the world hitherto almost unknown.

I stopped a night at San Jose, where I was most hospitably received by the alcalde, an American gentleman of intelligence, to whom I had a letter of introduction. Next day, after a pleasant ride of forty-five miles, I reached the Mission of San Juan, one of the most eligibly located of all the old missionary establishments. It was now in a state of decay. The vineyards were but partially cultivated, and the secos, or ditches for the irrigation of the land, were entirely dry. I got some very good pears from the old Spaniard in charge of the mission—a rare luxury after a long sea-voyage. The only tavern in the place was the "United States," kept by an American and his wife in an old adobe house, originally a part of the missionary establishment. Having secured accommodations for my mule, I took up my quarters for the night at the "United States." The woman seemed to be the principal manager. Perhaps I might have noticed her a little closely, since she was the only white woman I had enjoyed the opportunity of conversing with for some time. It was very certain, however, that she struck me as an uncommon person — tall, raw-boned, sharp, and masculine—with a wild and piercing expression of eye, and a smile singularly startling and unfeminine. I even fancied that her teeth were long and pointed, and that she resembled a picture of an ogress I had seen when a child. The man was a subdued and melancholy-looking person, presenting no particular trait of character in his appearance save that of general abandonment to the influence of misfortune. His dress and expression impressed me with the idea that he had experienced much trouble, without possessing that strong power of recuperation so common among American adventurers in California.

It would scarcely be worth while noticing these casual acquaintances of a night, since they have nothing to do with my narrative, but for the remarkable illustration they afford of the hardships that were encountered at that time on the emigrant routes to California. In the course of conversation with the man, I found that he and his wife were among the few survivors of a party whose terrible sufferings in the mountains during the past winter had been the theme of much comment in the newspapers. He did not state—what I already knew from the published narrative of their adventures—that the woman had subsisted for some time on the dead body of a child belonging to one of the party. It was said that the man had held out to the last, and refused to participate in this horrible feast of human flesh.

So strangely impressive was it to be brought in direct contact with a fellow-being, especially of the gentler sex, who had absolutely eaten of human flesh, that I could not but look upon this woman with a shudder. Her sufferings had been intense; that was evident from her marked and weather-beaten features. Doubtless she had struggled against the cravings of hunger as long as reason lasted. But still the one terrible act, whether the result of necessity or insanity, invested her with a repellant atmosphere of horror. Her very smile struck me as the gloating expression of a cannibal over human blood. In vain I struggled against this unchristian feeling. Was it right to judge a poor creature whose great misfortune was perhaps no offense against the laws of nature? She might be the tenderest and best of women—I knew nothing of her history. It was a pitiable case. But, after all, she had eaten of human flesh; there was no getting over that.

When I sat down to supper this woman was obliging enough to hand me a plate of meat. I was hungry, and tried to eat it. Every morsel seemed to stick in my throat. I could not feel quite sure that it was what it seemed to be. The odor even disgusted me. Nor could

I partake of the bread she passed to me with any more relish. It was probably made by her hands—the same hands that had torn the flesh from a corpse and passed the reeking shreds to her mouth. The taint of an imaginary corruption was upon it.

The room allotted to me for the night was roughly furnished, as might reasonably be expected; but, apart from this, the bedding was filthy; and, in common with every thing about the house, the slatternly appearance of the furniture did not tend to remove the unpleasant impression I had formed of my hostess. Whether owing to the vermin, or an unfounded suspicion that she might become hungry during the night, I slept but little. The picture of the terrible ogress that I had seen when a child, and the story of the little children which she had devoured, assumed a fearful reality, and became strangely mingled in my dreams with this woman's face. I was glad when daylight afforded me an excuse to get up and take a stroll in the fresh air.

CHAPTER II.

THE MIRAGE.

After an early breakfast, I mounted my mule and pursued my journey over the pass of the San Juan. The view from the summit was magnificent. Beyond a range of sand-hills toward the right stretched the great Pacific. Ridges of mountains, singularly varied in outline, swept down in front into the broad valley of the Salinas. The pine forests of Monterey and Santa Cruz were dimly perceptible in the distance; and to the left was a wilderness of rugged cliffs, as far as the eye could reach, weird and desolate as a Cape Horn sea suddenly petrified in the midst of a storm. Descending through a series of beautiful little valleys clothed in a golden drapery of wild oats, and charmingly diversified with groves of oak

PASS OF SAN JUAN.

and sycamore, and rich shrubbery of ceonosa, hazel, and wild grape, I at length entered the great valley of the Salinas, nine miles from the Mission of San Juan. At that time innumerable herds of cattle covered the rich pastures of this magnificent valley; and although there are still many to be seen there, the number has been greatly reduced during the last ten years. A large portion of the country bordering on the Salinas River, as far south as the Mission of Soledad, has been cut up into small ranches and farms; and thriving settlements and extensive fields of grain are now to be seen where formerly ranged wild bands of cattle, mustang, and innumerable herds of antelope.

Turning to the southward, and keeping in view the two great ranges of mountains which were the chief landmarks in former times, the scene that lay outspread before me resembled rather some wild region of enchantment than any thing that could be supposed to exist in a material world—so light and hazy were the distant mountains, so vaguely mingled the earth and sky, so rich and fanciful the atmospheric tints, and so visionary the groves that decorated the plain. Never before had I witnessed the mirage in the full perfection of its beauty. The whole scene was transformed into a series of magnificent optical illusions, surpassing the wildest dreams of romance. Points of woodland, sweeping from the base of the mountains far into the valley, were reflected in mystic lakes. Herds of cattle loomed up on the surface of the sleeping waters like miniature fleets of vessels with variegated sails. Mounds of yellow sand, rising a little above the level of the plain, had all the effect of rich Oriental cities, with gorgeous palaces of gold, mosques, and minarets, and wondrous temples glittering with jewels and precious stones. Bands of antelope coursed gracefully over the foreground; but so light and vaguely defined were their forms that they seemed rather to sail through the air than touch the earth. By the illusory process of the refraction, they appeared to sweep into the

ANTELOPE IN THE MIRAGE.

lakes and assume the forms of aerial boats, more fanciful and richly colored than the caïques of Constantinople. Birds, too, of snowy plumage, skimmed over the silvery waste; and islands that lay sleeping in the glowing light were covered with myriads of water-fowl. A solitary vulture, sitting upon the carcass of some dead animal a

VULTURE IN THE MIRAGE.

few hundred yards off, loomed into the form of a fabulous monster of olden times, with a gory head, and a beak that opened as if to swallow all within his reach. These wonderful features in the scene were continually changing: the lakes disappeared with their islands and fleets, and new lakes, with still stranger and more fantastic illusions, merged into existence out of the rarefied atmosphere. Thus hour after hour was I beguiled on my way through this mystic region of enchantment.

Toward evening I reached the Salinas River, where I stopped to rest and water my mule. A Spanish vaquero, whom I found under the trees enjoying the siesta to which that race are addicted, informed me that it was "*Dos leguos, poco mas o meno,*" to Soledad. As he lived there, he would show me the way. It was inhabited by the Sobranis family, and they owned sixteen square leagues of land and "*muchos granada.*" This much I contrived to understand; but when I handed the vaquero a fine Principe cigar, and he took a few whiffs and became eloquent, I entirely lost the train of his observations. It is possible he may have been reciting a poem on pastoral life. At all events, we jogged along very sociably, and in something over an hour reached the mission.

A more desolate place than Soledad can not well be imagined. The old church is partially in ruins, and the adobe huts built for the Indians are roofless, and the walls tumbled about in shapeless piles. Not a tree or shrub is to be seen any where in the vicinity. The ground is bare, like an open road, save in front of the main building (formerly occupied by the priests), where the carcasses and bones of cattle are scattered about, presenting a disgusting spectacle. But this is a common sight on the Spanish ranches. Too lazy to carry the meat very far, the rancheros generally do their butchering in front of the door, and leave the Indians and buzzards to dispose of the offal.

A young Spaniard, one of the proprietors, was the only person at home, with the exception of a few dirty Indians who were lying about the door. He received me rather coldly, as I thought, and took no concern

SOLEDAD.

whatever about my mule. I learned afterward that this family had been greatly imposed upon by travelers passing northward to the mines, who killed their cattle, stole their corn, stopped of nights and went away without paying any thing. At first they freely entertained all who came along in the genuine style of Spanish hospitality; but, not content with the kind treatment bestowed upon them, their rough guests seldom left the premises without carrying away whatever they could lay hands upon. This naturally embittered them against strangers, and of course I had to bear my share of the ill feeling manifested toward the traveling public. It was not long, however, before I discovered a key to my young host's good graces. He was strumming on an old guitar when I arrived, and soon resumed his solitary amusement, not seeming disposed to respond to my feeble attempts at his native language, but rather enjoying the idea of drawing himself into the doleful sphere of his own music. As soon as a favorable opportunity occurred, I took the guitar, and struck up such a lively song of "The Frogs that tried to Come it, but couldn't get a Chance," that the cadaverous visage of my host gradually relaxed into a smile, then into a broad grin, and at the climax he absolutely laughed. It was all right. Music had soothed the savage breast. Sobranis was conquered. He immediately directed the vaquero to see to my animal, and set to work and got me an excellent supper of tortillas and frijoles, jerked beef and oja; after which he insisted upon learning the song of the Frog, which of course I was obliged to teach him. So passed the hours till late bedtime. Notwithstanding the fleas, which abounded in overwhelming numbers, I contrived to sleep soundly. Next morning, after a good breakfast of coffee, tortillas, jerked beef, etc., as before, I mounted my mule and proceeded on my journey, much to the regret of Sobranis, who positively refused to accept a cent for the accommodations he had afforded me.

CHAPTER III.

A DEATH-STRUGGLE.

In the vicinity of the sea-shore, and as far inland as Soledad, the temperature was delightfully cool and bracing; but beyond the first turning-point of mountains to the southward a marked change was perceptible. Although the sun was not more than two hours high, the heat was intense. The rich black soil, which had been thoroughly saturated with the winter rains, was now baked nearly as hard as stone, and was cracked open in deep fissures, rendering the trail in some places quite difficult even for the practiced feet of the mule. Every thing like vegetation was parched to a crisp with the scorching rays of the sun. The bed of the river was quite dry, and no sign of moisture was visible for many miles. The rich fields of wild oats were no longer to be seen, but dried and cracking wastes of wild mustard, sage-weed, and bunch grass. In some places deserts of sand, without a particle of vegetation, and incrusted with saline deposits, stretched along the base of the mountains as far as the eye could reach. The glare on these plains of alkali (as they were commonly called) was absolutely blinding. Toward noon, so intense was the heat, I thought it impossible to endure it another hour. A dry, hot cloud of dust rose from the parched earth, and hung around me like the fiery breath of an oven. Neither tree nor shrub was to be seen any where along the wayside. As I toiled wearily along, scarcely able to get my mule out of a walk, I thought of Denham and Clapperton, the brothers Lander, Mungo Park, and all the great African explorers, and wondered how they could have endured for weeks and months what I found it so hard to bear for a few hours. There was no respite; nothing in the world to alleviate the burning heat; not even a stunted shrub to creep under. And yet, thought I, this is but a flash in the pan to the deserts of Africa. Not that the heat is more intense there; for I believe it is admitted that the thermometer rises higher in California than in any other part of the world. I have known it to be 130° Fahrenheit in the mines, and have been told that in the gulches of some of the foothills of the Sierra Nevada it has been known to reach 150°. The official table published by Congress shows that the maximum heat at Fort Miller is 118°, while at Fort Yuma, on the Colorado, it does not exceed 110°. In the narrative of the voyages of Lord Anson, written by his chaplain, it is conceded that the heat is greater in California, owing to local causes, than at any known point between the tropics. But very different is it in Africa, or any tropical country, in this respect—that the climate of California is never oppressive, whatever may be the temperature. The nights are delightfully cool, and the mornings peculiarly fresh and bracing. Hence the suffering from heat is never protracted beyond a few hours. At all events, not to go into any farther dissertation upon climate, I found it quite warm enough on the present occasion, and would have been very glad to accept the loan of an umbrella had any body been at hand to offer it to me.

About an hour before sunset, as I was riding slowly along, enjoying the approaching shades of evening, I discovered for the first time that my mule was lame. I had traveled very leisurely on account of the heat, making not over thirty miles. The nearest water, as the young Spaniard, Sobranis, had informed me, was at a point yet distant about five miles. I saw that it was necessary to hurry, and began to spur my mule in the hope of being able to reach this camping place; but I soon perceived that the poor animal was not only lame, but badly foundered—at least it seemed so then, though my convictions on that point were somewhat shaken by what subsequently occurred: I had succeeded, after considerable spurring, in getting him into a lope, when he suddenly stumbled and threw me over his head. The shock of the fall stunned me for a few moments, but fortunately I was not hurt. I must have turned a complete somersault. As soon as consciousness returned I found that I was lying on my back in the middle of the road, the mule quietly grazing within ten feet. I got up a little bewildered, shook off some of the dust, and started to regain the bridle; but, to my great surprise, the mule put back his ears, kicked up his heels, and ran off at a rate of speed that I deemed a foundered animal entirely incapable of achieving. There was not the slightest symptom of lameness in his gait. He "loped" as freely as if he had just begun his journey. In vain I shouted and ran after him. Sometimes he seemed absolutely to enjoy my helpless condition, and would permit me to approach within two or three feet, but never to get hold of the bridle. Every attempt of that kind he resented by whirling suddenly and kicking at me with both heels, so that once or twice it was a miracle how I escaped. For the first time since morning, notwithstanding the heat of the day, my skin became moist. A profuse sweat broke out all over me, and I was parched with a burning thirst. It was thirty miles from Soledad, the nearest inhabited place that I knew of, and even if I felt disposed to turn back it would have been at great risk and inconvenience. My blankets, coat, pistol, and papers—the whole of incalculable importance to me—were firmly strapped behind the saddle, and there was no way of getting at them without securing the mule. Upon reflection, it seemed best to follow him to the watering-place. He must be pretty thirsty after his hard day's journey in the sun, and would not be likely to pass that. I therefore walked on as fast as possible, keeping the mule as near in the trail as his stubborn nature would permit. It was not without difficulty, however, that I could discern the right trail, for it was frequently intersected by others, and occasionally became lost in patches of sand and sage-brush.

In this way, with considerable toil, I had advanced about two miles, when I discovered that a large band of Spanish cattle, which had been visible for some time in the distance, began to close in toward the line of my route, evidently with the intention of cutting me off. Their gestures were quite hostile enough to inspire a solitary and unarmed footman with uneasiness. A fierce-looking bull led the way, followed by a lowing regiment of stags, steers, and cows, crowding one upon the other in their furious charge. As they advanced, the leader occasionally stopped to tear up the earth and shake his horns; but the mass kept crowding on, their tails switching high in the air, and uttering the most fearful bellowing, while they tossed their horns and stared wildly, as if in mingled rage and astonishment. I had heard too much of the wild cattle of California, and their hostility toward men on foot at this season of the year, not to become at once sensible of my dangerous position.

The nearest tree was half a mile to the left, on the margin of a dry creek. There was a grove of small oaks winding for some distance along the banks of the creek; but between the spot where I stood and this place of security scattering bands of cattle were grazing. However, there was no time to hesitate upon a choice of difficulties. Two or three hundred wild cattle rushing furiously toward one in an open plain assist him in coming to a very rapid conclusion. I know of no position in which human strength is of so little avail—the tremendous aggregation of brute force opposed to one feeble pair of arms seems so utterly irresistible. I confess instinct lent me a helping hand in this emergency. Scarcely conscious of the act, I ran with all my might for the nearest tree. The thundering of heavy hoofs after me, and the furious bellowing that resounded over the plain, spread a contagion among the grazing herds on the way, and with one accord they joined in the chase. It is in no spirit of boastfulness that I assert the fact, but I certainly made that half mile in as few minutes as ever the same distance was made by mortal man. When I reached the tree I looked back. The advance body of the cattle were within a hundred yards, bearing down in a whirlwind of dust. I lost no time in making my retreat secure. As the enemy rushed in, tearing up the earth and glaring at me with their fierce, wild eyes, I had gained the fork of the tree, about six feet from the ground, and felt very thankful that I was beyond their reach. Still there was something fearful in being blockaded in such a place for the night. An intolerable thirst parched my throat. The effects of the exertion were scarcely perceptible at first, but as I regained my breath it seemed impossible to exist an hour longer without water. In this valley the climate is so intensely dry during the summer heats that the juices of the system are quickly absorbed, and the skin becomes like a sheet of parchment. My head felt as if compressed in a band of iron; my tongue was dry and swollen. I would have given all I possessed, or ever hoped to possess, for a single glass of water.

While in this position, with the prospect of a dreary night before me, and suffering the keenest physical anguish, a very singular circumstance occurred to relieve me of farther apprehension respecting the cattle, though it suggested a new danger for which I was equally unprepared. A fine young bull had descended the bed of the creek in search of a water-hole. While pushing his way through the bushes he was suddenly attacked by a grizzly bear. The struggle was terrific. I could see the tops of the bushes sway violently to and fro, and hear the heavy crash of drift-wood as the two powerful animals writhed in their fierce embrace. A cloud of dust rose from the spot. It was not distant over a hundred yards from the tree in which I had taken refuge. Scarcely two minutes elapsed before the bull broke through the bushes. His head was covered with blood, and great flakes of flesh hung from his fore shoulders; but, instead of manifesting signs of defeat, he seemed literally to glow with defiant rage. Instinct had taught him to seek an open space. A more splendid specimen of an animal I never saw; lithe and wiry, yet wonderfully massive about the shoulders, combining the rarest qualities of strength and symmetry. For a moment he stood glaring at the bushes, his head erect, his eyes flashing, his nostrils distended, and his whole form fixed and rigid. But scarcely had I time to glance at him when a huge bear, the largest and most formidable I ever saw in a wild state, broke through the opening.

A trial of brute force that baffles description now ensued. Badly as I had been treated by the cattle, my sympathies were greatly in favor of the bull, which seemed to me to be much the nobler animal of the two. He did not wait to meet the charge, but, lowering his head, boldly rushed upon his savage adversary. The grizzly was active and wary. He no sooner got within reach of the bull's horns than he seized them in his powerful grasp, keeping the head to the ground by main strength and the tremendous weight of his body, while he bit at the nose with his teeth, and raked stripes of flesh from the shoulders with his hind paws. The two animals must have been of very nearly equal weight. On the one side there was the advantage of superior agility and two sets of weapons—the teeth and claws; but on the other, greater powers of endurance and more inflexible courage. The position thus assumed was maintained for some time—the bull struggling desperately to free his head, while the blood streamed from his nostrils—the bear straining every muscle to drag him to the ground. No advantage seemed to be gained on either side. The result of the battle evidently depended on the merest accident.

As if by mutual consent, each gradually ceased struggling, to regain breath, and as much as five minutes must have elapsed while they were locked in this motionless

A DUEL À LA MORT.

but terrible embrace. Suddenly the bull, by one desperate effort, wrenched his head from the grasp of his adversary, and retreated a few steps. The bear stood up to receive him. I now watched with breathless interest, for it was evident that each animal had staked his life upon the issue of the conflict. The cattle from the surrounding plains had crowded in, and stood moaning and bellowing around the combatants; but, as if withheld by terror, none seemed disposed to interfere. Rendered furious by his wounds, the bull now gathered up all his energies, and charged with such impetuous force and ferocity that the bear, despite the most terrific blows with his paws, rolled over in the dust, vainly struggling to defend himself. The lunges and thrusts of the former were perfectly furious. At length, by a sudden and well-directed motion of his head, he got one of his horns under the bear's belly, and gave it a rip that brought out a clotted mass of entrails. It was apparent that the battle must soon end. Both were grievously wounded, and neither could last much longer. The ground was torn up and covered with blood for some distance around, and the panting of the struggling animals became each moment heavier and quicker. Maimed and gory, they fought with the desperate certainty of death —the bear rolling over and over, vainly striking out to avoid the fatal horns of his adversary—the bull ripping, thrusting, and tearing with irresistible ferocity.

At length, as if determined to end the conflict, the bull drew back, lowered his head, and made one tremendous charge; but, blinded by the blood that trickled down his forehead, he missed his mark, and rolled headlong on the ground. In an instant the bear whirled and was upon him. Thoroughly invigorated by the prospect of a speedy victory, he tore the flesh in huge masses from the ribs of his prostrate foe. The two rolled over and over in the terrible death-struggle; nothing was now to be seen save a heaving, gory mass, dimly perceptible through the dust. A few minutes would certainly have terminated the bloody strife, so far as my favorite was concerned, when, to my astonishment, I saw the bear relax in his efforts, roll over from the body of his prostrate foe, and drag himself feebly a few yards from the spot. His entrails had burst entirely through the wound in his belly, and now lay in long strings over the ground. The next moment the bull was on his legs, erect and fierce as ever. Shaking the blood from his eyes, he looked around, and seeing the reeking mass before him, lowered his head for the final and most desperate charge. In the death-struggle that ensued both animals seemed animated by supernatural strength. The grizzly struck out wildly, but with such destructive energy that the bull, upon drawing back his head, presented a horrible and ghastly spectacle; his tongue, a mangled mass of shreds, hanging from his mouth, his eyes torn completely from their sockets, and his whole face stripped to the bone. On the other hand, the bear was ripped completely open, and writhing in his last agonies. Here it was that indomitable courage prevailed; for, blinded and maimed as he was, the bull, after a momentary pause to regain his wind, dashed wildly at his adversary again, determined to be victorious even in death. A terrific roar escaped from the dying grizzly. With a last frantic effort he sought to make his escape, scrambling over and over in the dust. But his strength was gone. A few more thrusts from the savage victor, and he lay stretched upon the sand, his muscles quivering convulsively, his huge body a resistless mass. A clutching motion of the claws —a groan—a gurgle of the throat, and he was dead.

The bull now raised his bloody crest, uttered a deep bellowing sound, shook his horns triumphantly, and slowly walked off, not, however, without turning every few steps to renew the struggle if necessary. But his last battle was fought. As the blood streamed from his wounds a death-chill came over him. He stood for some time, unyielding to the last, bracing himself up, his legs apart, his head gradually drooping; then dropped on his fore knees and lay down; soon his head rested upon the ground; his body became motionless; a groan, a few convulsive respirations, and he too, the noble victor, was dead.

During this strange and sanguinary struggle, the cattle, as I stated before, had gathered in around the combatants. The most daring, as if drawn toward the spot by the smell of blood or some irresistible fascination, formed a circle within twenty or thirty yards, and gazed at the murderous work that was going on with startled and terror-stricken eyes; but none dared to join in the defense of their champion. No sooner was the battle ended, and the victor and the vanquished stretched dead upon the ground, than a panic seized upon the excited multitude, and by one accord they set up a wild bellowing, switched their tails in the air, and started off at full speed for the plains.

CHAPTER IV.

THE OUTLAWS' CAMP.

It was now nearly dark. The impressive scene I had just witnessed, and in which I had become so absorbed as to lose all consciousness of danger, now forcibly reminded me that this was not a safe place of retreat for the night. I descended from the tree, seeing all clear, and hurried out toward the edge of the plain, where I discovered a trail leading down parallel with the creek. The water-hole I knew must be on this creek, for there was no other in sight. It could not be more than two or three miles distant, and there was yet sufficient light to enable me to keep within range of the bushes on the left. I walked on rapidly for nearly an hour, sometimes stumbling into the deep fissures which had been made in the ground by the heat of the sun, and often obliged to descend deep arroyas and seek for some time before I could find an outlet on the other side; but in the course of an hour I was rejoiced to see a point of woodland jutting into the plain, not over a few hundred yards distant, in the midst of which there was the glimmer of a fire.

I say rejoiced, for certainly that was the first sensation; but in approaching the light I could not but think of the savage character of the country, and the probability of meeting with company here as little to my liking as any I had yet encountered. This part of the Salinas was entirely out of the range of civilization; neither miners nor settlers had yet intruded upon these dreary solitudes; and the chances were greatly in favor of meeting a party of Sonoranian desperadoes or outlawed Californians. Yet what inducement could I present for robbery or murder in such a destitute plight? Without coat, blankets, pistol, or property of any kind except a watch concealed in the fob of my pantaloons—even without money; for what little I owned, not over forty or fifty dollars, was contained in a leather purse in the pocket of my coat—of what avail would it be to molest me? If plunder should be an object, they must already be in possession of all I had.

These considerations somewhat allayed my apprehensions; and, at all events, I saw no alternative but to keep on. As I descended from the plain into the oak grove bordering upon the bed of the creek, I observed that there were only two men in camp. From their costume—the common blue shirts, pantaloons, and rough boots of ordinary travelers on the way to the mines—I judged them to be Americans. Nor was I mistaken. The very first word I heard spoken was an oath, which it is unnecessary for me to repeat.

"I say, Griff," said one, in a coarse, brutal voice, "if he comes don't you budge. He'll be here certain."

"Jack," replied the man addressed, "you've done enough of that. You'd better hold up a while, that's my opinion."

The other laughed; not a joyous laugh of natural mirthfulness, but something resembling a chuckling sneer that was horribly repelling. An instinctive feeling prompted me to retrace my steps and strike out for the Mission of Soledad. Without well knowing why, I was impressed with an irresistible conviction that the spirit of sin brooded over this camp. Acting upon the impulse of the moment, I turned to retreat while yet undiscovered, when a man emerged from the bushes a little below, and called out sharply, "Who's that? Answer quick, or you're a dead man!"

I answered at once, "An American—a friend. Don't shoot! It's all right!"

I then advanced into the camp, where I was greeted with an uneasy and suspicious stare, very much unlike any reception I had ever met with before from a party of countrymen. There was either distrust or disappointment in their looks, probably both. The party consisted of three men, two of whom were standing by the fire cooking a piece of venison, while the third, who had hailed me from the bushes, seemed to have been on the look-out.

The man called "Jack"—he who had first spoken—was a swarthy, thick-set fellow, about thirty years of age, with a bull neck, a coarse black beard, and heavy sun-burned mustache. His eyes were overhung by bushy brows, and were of a cold, stony color and very deeply set, giving him an appearance of peeping out furtively from a chaparral of brush. A shock of black matted hair covered his head; his hands were begrimed with dirt, and his dress was ragged, greasy, and stained with blotches of filth and blood. On his feet he wore a pair of coarse heavy boots, out at the toes, in the legs of which his pantaloons were carelessly thrust, giving him a peculiarly slovenly and blackguard air. A belt around his waist, with a revolver and knife, and a leather pouch for balls and patching, completed his costume and trappings. I instinctively recoiled from this man. His

THE CAMP.

"JACK."

whole expression—his voice, manner, dress, and all—pronounced him a coarse and unmitigated villain. There was not a single redeeming point about him that I could discover. Hard, crafty, and cruel, profane, filthy, and brutal, his character was patent at a glance. If he was not intrinsically bad, naturo had grievously belied him.

The other, to whom this fellow had addressed his remarks when I first heard their voices, and who was called "Griff," was apparently somewhat younger, though rough and weather-beaten, as if he had been much exposed. His form was gaunt and athletic, and his height over six feet. There was something very sad in the expression of his face, which was well chiseled, and not destitute of a certain quality of rough, manly beauty. A prominent nose; firm and compressed lips; a square projecting chin, evincing firmness, and a liquid blue eye, with a mingled expression of gentleness and determination; deep furrows, tending downward from the corners of his mouth; long waving hair, and a light mustache, gave him something of a heroic cast of countenance, which, but for an appearance of general recklessness, would have redeemed him under all the disadvantages of ragged clothes and evil associations. Yet I felt at once interested in this man. He seemed embarrassed as I scanned his features, apparently struggling with some natural impulse of politeness, which prompted him to offer me a more kindly welcome than his comrades had bestowed upon me; but, if such an impulse moved him, it was speedily checked. He drew his hat over his brow, and resumed his occupation at the fire without saying a word. Still, even his silence was not unfriendly.

The third of this strange party was a lithe, wiry man, not over five feet eight in height, but compact and not ungracefully formed. He was apparently much older than either of the others. To look upon him once was to receive an impression of evil that could never be effaced. His countenance was the most repellent I had ever seen, far surpassing that of the man "Jack" in cool, crafty malignity. I could readily imagine that this was the leader in all that required subtlety, intellect, and skill. His forehead was high and narrow; his eyes closely set together, black, and of piercing brilliancy; his features sharp and mobile; but it was his mouth that more than all gave him the distinguishing expression of cruelty and cunning. A sardonic smile continually played upon his thin, bloodless lips. Every muscle seemed under perfect control. It might well be said of this man that

"He could smile, and smile, and be a villain still,"

for villainy lurked in every feature. Yet he was not deficient in a certain air of personal neatness to which the other two had no pretensions. His jet-black hair was closely cut, and his face quite destitute of beard, and of that peculiar leaden color which indicates a long career of dissipation. In his dress he was even slightly foppish; wore a green cassimere hunting-jacket, with brass buttons; a white shirt, a breast-pin, and a pair of check pantaloons. His fingers were adorned with rings, and a watch-guard hung from his neck. The hilt of a bowie-knife, ornamented with silver, protruded from under the breast of his vest, and a revolver hung from a belt around his waist. In his motions he was quick, supine, and noiseless. Something of the basilisk there was about this man—something brilliant and glossy, as if he shone with a peculiar light. I fancied I had seen gamblers like him in New Orleans, fierce yet wary men, accustomed to play at hazardous games; glossy outside and of fascinating suavity, but corrupt to the core. Even his green coat added to the illusion; it fitted him so neatly, and seemed so like the natural slimy skin of a poisonous reptile. It was evident this was no ordinary adventurer. His manner was that of a man of the world; he had seen much, and he knew much, mostly of evil I fancied, for all that was about him was essentially bad. A certain deference toward him was perceptible in the manner of the other two men, especially in that of the thick-set fellow called Jack, who lost much of his bravado air when "the Colonel" spoke, for such was the title accorded to the last-named of the party. The Colonel was pleased to scan me very closely for some moments before he opened his lips. When he spoke I was astonished at the change in his voice, which, when I first heard it, was sharp and hard. It was now wonderfully soft and silky.

"Sir," said he, blandly, "you seem to have lost your way. Have you walked far?"

"Not very," was my answer. "Only five miles. My mule threw me and ran away. I was unable to catch him, and thought probably he had made his way to this pool of water. Have you seen him?—a large brown mule, with a roll of blankets and a coat fastened to the saddle?"

The Colonel smiled pleasantly.

"I see, friend, you are not accustomed to traveling in this rough style. Your mule has doubtless gone back to his old quarters, wherever you got him. A mule never goes farther in a new direction than he can help."

"But I saw him start for this point. He was very thirsty, I know; and, besides, he came from the Colorado not over a month ago. His course would naturally be to the southward if he desired to return to his old quarters."

"Very likely," said the Colonel, quietly: "it may be the same mule I sold to a gentleman from Texas down there about that time."

"Yes—I bought him from a Texan. It must be the same," I answered, glad to find some clew, however remote, to the object of my search.

The Colonel smiled again, and expressed his regret that it was not the nature of that mule to go in the direction of the Colorado. The fare for mules in that region was rather dry; and the animal in question had a very keen appreciation of good fare. At all events, no such mule had been seen here—"unless, perhaps, you may have seen him," added the Colonel, turning to the thick-set man, and regarding him with a peculiar expression—the same basilisk eye that I had noticed before.

"I?" said Jack, laughing coarsely; "the last mule I saw was a small mustang horse that belongs to myself."

"Possibly *you* may have seen him?" suggested the Colonel, looking at the tall, gaunt man, Griff; and here I could not but notice the change in his expression. His brow unconsciously lowered, and there was something devilish in the cool malignity of his eye. Griff was silent. His frame seemed convulsed with some emotion of disgust or hatred. The Colonel, turning quickly to me, observed, with an affected suavity, "This man may possibly be able to tell you something about your mule."

At this the person referred to drew himself up into an erect position, and gave a look at the Colonel—a look of such mingled hatred, defiance, and contempt, that I expected to see the latter wilt before it or draw his revolver. But he did neither. And here I detected the secret of his power over the other two men—imperturbable self-possession. He merely elevated his brows superciliously as Griff sternly remarked,

"You know as much of the mule as I do! What do you ask me for? Be careful."

"Oh," said the Colonel, jocularly, "I thought you might have seen him while I was absent. You know I'm not in the habit of noticing these things."

Griff resumed his slouching attitude, stirring the fire moodily, while the Colonel requested me to be seated, and proceeded to do the honors of the repast. All that I have attempted to describe was perfectly quiet; not a loud word was spoken, and but for the peculiar expression of each face, involving some dark complicity of experience, it might have passed unnoticed. There was really nothing said that necessarily bore an evil import. Yet what was it that filled me with such an indefinable abhorrence of these men—of two of them, at least? That they were unprincipled adventurers, I knew; that they were depraved enough to be professed gamblers, highway robbers, or horse thieves, was reasonable to suppose from their appearance; but there was something more than that about them. The leader was no common gambler or horse-thief. He was too keen, too polished, too subtle for that. He might be a forger, a slave speculator, a dealer in blood-hounds, a gambler in fancy stocks; yet this was no country for the exercise of that sort of talent—at least that portion of it which he had chosen as a place of temporary abode. He might be on his way to the mines. I asked no questions. It was enough to feel the evil influence of the present—enough to know by intuition that the hands of this man were stained with some deadly sin.

Hungry as I was, I could not swallow the bread he gave me without a choking sensation of disgust. The act of eating with him implied a species of fellowship against which my very soul rebelled.

Of the swarthy man, Jack, I had a different impression. He was purely brutal. All his instincts were coarse, savage, and depraved. Whatever quickness or cunning he possessed was that of an animal. He was far inferior to the other in all the essential attributes of a successful villain. I looked upon him as upon a vicious brute.

For the tall fellow, Griff, I must confess I felt a strange sympathy. That he was not naturally depraved, no one who looked upon his fine features, and frank, manly bearing, could for a moment doubt. He might be dissipated, reckless, even criminal, but he surely was not all bad. There was something of conscience left in him yet — some human emotion of remorse. Otherwise, why was his expression so strangely sad? Why was it that there seemed to be no bond of sympathy between him and the others — beyond, perhaps, some complicity in crime, either accidental or the result of evil associations? A deadly fascination seemed to be spread over him by the leader, against which he struggled in vain. The slight outburst of passion which I had witnessed showed too plainly the powerful thraldom in which he was held. His defiant tone—the withering hatred of his eye—the impatient gesture of contempt, were but the momentary ebullitions of a proud spirit. No sentiment of personal fear could have found a place in that manly breast. The cause of his submission lay deeper than that. Something of self-accusation must have had a share in it, thus to paralyze his strength—something more inextricable than any web that mortal man could cast over him unaided by a sense of his own iniquity. I could not conjecture what crime he had committed. Whatever it was, I had a strong yearning to befriend him. Surely there was still hope for him; he could not be utterly lost without bearing in his features the impress of unmitigated evil.

As soon as supper was over, the Colonel lighted his pipe and seemed disposed to be sociable. It was impossible for me to get over the abhorrence I had for this man. Even his efforts to be agreeable had something sinister in them that increased my dislike. Still, I was in the power of these men, whether they chose to exercise it for good or for evil, and it behooved me to suppress any disrelish I might have for their company.

"You came from Soledad to-day, I think you said?" observed the Colonel.

"Yes; I stopped there last night."

"Did you meet any body on the road?" he asked, carelessly.

"Only two Spaniards from Santa Marguerita." The Colonel started.

"Any news from below?"

"None that I could understand. I don't speak the Spanish language."

"You heard nothing from San Miguel?"

"No."

"Which way are you bound, if I may take the liberty of asking?"

"To San Luis. I have business there connected with the revenue service. Unfortunately, my mule has disappeared with my blankets, coat, pistol, what little money I had, and my official papers, which are of no use to any body but myself. I fear the loss will subject me to great inconvenience."

"You are aware, I suppose," said the Colonel, with the same disagreeable smile I had before noticed, "that the road is considered a little dangerous for solitary travelers. Murders have been committed between this and San Miguel."

"Any lately?" I asked, assuming more composure than I felt.

"Why as for that," replied the Colonel, making an effort to be humorous, "it would be hard for me to keep the run of all I hear in this part of the country. Society is rather backward, and the newspapers do not keep us advised of the current events of the day."

Here there was a pause. I felt convinced that this man was capable of any deed, however dark and damning. Even while he spoke his fingers played with the butt of a revolver that hung from his belt. Something caught my eye as his hand moved—a small silver star near the lock of the pistol. This was not an ordinary mark. I at once knew the pistol to be mine. A friend had given it to me. The star was a fanciful device of his own, based upon the idea that its rays would guide the bullet to its destination. The Colonel detected my inquisitive glance, and smiled again in his peculiar way, but said nothing. If I had any doubt on the subject before, I now felt quite satisfied that he was not only a villain, but one who would not hesitate to take my life if it would serve his purpose. Whether his thoughts ran in that direction at present I could not determine. He possessed a wonderful power of inspiring dark impressions without uttering a word. The mere suspicion of such a design was at least unpleasant. At length he rose, having finished smoking his pipe, and with an air of indifference said,

"It must be getting late. Have you the time, sir?"

I pulled out my watch, scarcely conscious of the act, and remarked that it wanted a few minutes of nine.

"A nice-looking watch, that!" observed the Colonel. "It must be worth a hundred dollars."

"Yes, more than that," I answered; for I saw at once that any manifestation of suspicion would be the last thing to answer my purpose. "It cost $150 in New York. It is a genuine chronometer, and the casing is of solid gold."

The Colonel exchanged glances with the swarthy man, Jack, and proposed to go out and take a look at the horses. Before they had proceeded fifty yards they stopped and looked back. Griff had been sitting moodily before the fire during the conversation above related, and did not seem disposed to move at the summons of his leader, who now called sharply to him to come on. The same expression of defiant hatred that I had noticed before flashed from the man's eyes, and for a moment he seemed to struggle against the Colonel's malign influence. "Come!" said the latter, sharply, "what do you lag behind for? You know your duty!"

"Yes," muttered Griff, between his set teeth, "I know it! It is hardly necessary to remind me of it." He then rose and proceeded to join his comrades. As he passed by where I sat he hurriedly whispered, "*Stay where you are. Don't attempt to escape yet. Depend upon me—I'll stand by you!*"

CHAPTER V.

THE ESCAPE.

It may readily be conceived that my sensations were not the most pleasant during the absence of the three men in whose power I was so strangely and unexpectedly placed. That two of them were quite capable of murdering me, if they had not already made up their minds to do so, was beyond question. I looked around, and saw to my dismay that they scarcely took the trouble to conceal the robbery they had already perpetrated. My blankets lay under a tree not over fifteen steps from the fire, and my coat and saddle were carelessly thrown among the common camp equipments in the same place. What could one unarmed man do against three, or even two, fully armed desperadoes? My first impulse was to steal away, now that there was a chance—perhaps the only one I might have—and conceal myself in the bushes till morning, then endeavor to make my way along the bed of the creek to Soledad. Better trust to the grizzly bears than to such men as the Colonel and Jack. But it was more than probable they were thoroughly acquainted with every thicket and trail in the country, and would not be long in overtaking me on horseback. There was another serious consideration: I could not well afford to lose my mule, money, and papers. The latter were of incalculable value, and could not be replaced. I had no idea that they had been suffered to remain in my coat pocket. So adroit a speculator as the Colonel must have ascertained their contents and placed them beyond danger of recovery. Besides, the man Griff had warned me not to attempt an escape yet. Was he to be trusted? Surely I could not be deceived in him. What object could he have in warning me unless to provide for my safety?

These considerations were unanswerable. I determined to remain and abide the issue.

It is said that danger sharpens men's wits. I believe it; for while there was ample reason to suppose these men were deliberating upon my destruction, a scheme flashed upon my mind which I at once resolved to carry into effect. Up to this period I had given them a plain statement of my misfortune. They evidently regarded me as a very simple-minded and inexperienced traveler. Nothing could be easier than to improve upon that idea.

As soon as they returned and resumed their places around the fire, I made some casual inquiries of the Colonel about the route from San Miguel to San Luis Obispo, professing to be exceedingly anxious to reach the latter place within five or six days.

The Colonel was bland and obliging as usual, giving me, without reserve, full particulars in regard to the route.

"But what's your hurry?" said he, smiling in his accustomed manner; "why not stay with us a few days and make yourself comfortable? The weather is rather warm for so long a pedestrian tour — unless, indeed, something is to be made by it." This he said with a low chuckle and a significant glance at the fellow with the thick neck.

"That is precisely why I want to get on," I answered; "a great deal is to be made by it if I get there in time, and a great deal lost if I don't. A vessel laden with foreign goods has gone ashore on the beach below the Embarcadera. I have advices that most of the cargo is saved. The duties, according to a copy of the manifest forwarded to the Custom-house at San Francisco, amount to over ten thousand dollars. The supercargo writes that he can sell out on advantageous terms at San Luis, provided he can pay the duties there to some authorized officer of the government within the period named. I am on my way down to receive the money. If I can get back with it to San Francisco within ten or twelve days, it will be of considerable advantage to the government as well as to myself. Unfortunately, there is no water communication at present, or I might gain time by taking a vessel. However, I apprehend no difficulty in being able to hire a mule at San Miguel. As for the stories of robbery and murder on the road, I have no faith in them. At all events, I am not afraid to try the experiment."

This communication made an evident impression upon the minds of the Colonel and Jack, both of whom listened with intense interest. The man Griff looked a little puzzled, but a casual glance reassured him: he at once caught at my meaning. I could see that the Colonel was embarrassed as to what course to pursue in reference to the stolen property. He held down his head for some time, pretending to be occupied in clearing the stem of his pipe, but it was apparent that he was in considerable perplexity. Deep and guarded as he was, it was not difficult to conjecture what was passing in his mind. There was now a strong inducement for permitting me to proceed on my journey. The prospect of securing ten thousand dollars was worthy of some risk; yet, if he acknowledged the stealing of my mule and other property, it was not likely I would again place myself in his power. On the other hand, I had seen the pistol, and must have some suspicion of the true state of the case.

I have often observed that men deeply versed in villainy, while they possess a certain sort of sagacity, are deficient in the perception of character when it involves a more comprehensive knowledge of human nature than usually falls within the limits of their individual experience. They are quick to detect every species of vulgar trickery, but their capacity to cope with straightforward truth is limited. They suspect either too much or too little, and lose confidence in their own penetration. With men like themselves they understand how to deal—they know by intuition the governing motives; but simplicity and frankness are weapons to which they are not accustomed. A direct statement of facts, in which they can see no motive of prudence, sets them at fault. They can analyze well through a dark atmosphere, but, like night-birds, have very dim perceptive powers in daylight.

While the Colonel could discover no interested motive in my simple statement respecting the loss of a vessel on the coast (of which he had probably heard from other sources), and could see no reason why I should not be simple enough to come back with a large sum of money, since I had been simple enough to lose a valuable mule and exhibit a valuable watch, he nevertheless seemed unable to extricate himself from suspicion in reference to the pistol—the only article of my property which he had reason to suppose I had seen. He could easily have said that he had found it on the trail; but he was not skilled in degrees of innocence. He had deferred his explanation too long, and, judging by himself, could not imagine that any other person would credit so flimsy a statement. In this he was correct, but his one-sided sagacity led him into puzzling inconsistencies.

To lull all suspicion on this point was indispensable to the success of my plan. The apparent confidence which I had manifested in the good faith of the party tended greatly to prevent the leader from coming to a satisfactory conclusion. So at least it appeared to me, as I watched the uncertain movements of his hands and the changing expression of his countenance. He was evidently aware that I had seen the star on the handle of the pistol, yet my conduct indicated no suspicion. It was necessary that I should remove whatever doubt on the subject might be lurking in his mind. With this in view, I took occasion to renew the conversation relative to the route, stating that although I apprehended little danger, it was still an awkward position to be entirely without arms in a strange country.

"The loss of my pistol," said I, "is a serious inconvenience. It must have fallen from my belt when the mule threw me, and become covered with dust. I could go back and find the place, but that would occupy nearly half a day, and I can not afford to lose the time. The only particular value the pistol has is that it is a present from a friend who belonged to the Order of the Lone Star of Texas. The badge of the Association is marked upon the handle, as usual with arms belonging to the members."

"Yes," said the Colonel, after a pause, "I once belonged to that Order myself, and have a pistol similarly marked."

"Perhaps you would be willing to dispose of it?" I observed. "Not that I have any money, but I would cheerfully give my watch for a good pistol, which would be at least three times its value."

"My dear sir," said the Colonel, affecting an air of injured pride, "you certainly can not be aware that a member of the Lone Star never sells or barters his arms. Any thing else, but not his weapons of personal defense. Fortunately, however, I have a spare revolver, which is entirely at your service. As for your watch, I should be sorry to deprive you of so useful an article, and one which would be of no value to myself. Time is of little consequence to men who are accustomed to spend it as they please, and whose chief dependence is on the sun, moon, and stars."

I accepted the proffered gift, as may be supposed, without the slightest qualms of conscience in depriving the donor of so valuable a piece of property; and having expressed my thanks, noticed that, while pretending to search for the pistol among the camp equipments, he took care to cover up my blanket and coat.

The Colonel soon returned to the fire, and handed me a very handsome revolver, a belt, powder-flask, and small leather bag containing caps, balls, and other necessary appendages. It struck me as a little strange that, having apparently made up his mind to let me depart, he had not offered to lend me an animal to ride upon; but a moment's reflection satisfied me that there was good cause for this. There could be no doubt, from the character of the party, that the horses were stolen, and would be recognized on the road. Besides, he knew I could easily hire a horse or mule at San Miguel.

After this I observed that the Colonel took occasion to speak a few words to Jack, the import of which I could only conjecture had some reference to my papers. Jack answered aloud, "Yes, the grass is bad there. I'll go put my mustang in another place." He then walked away, and the Colonel busied himself in preparing our sleeping quarters for the night.

It was nearly eleven o'clock. In about fifteen minutes Jack returned, and we all lay down in different directions, within a short distance of the fire. A saddle-blanket, kindly furnished by my chief entertainer, enabled me to make quite a comfortable bed.

The night was mild and pleasant. A clear sky, spangled with stars, was visible through the tops of the trees, and never had I seen it look so beautifully serene. Could it be that guilt could slumber peacefully under that heavenly canopy? Surely the evil spirit must be strong in the hearts of men who, unconscious of the reproving purity of such a night, could thus forget their sins, and lie calmly sleeping upon the bosom of their mother earth. How deadened by a long career of crime must conscience be in the breast of him who, steeped in guilt, could thus, in the presence of his Maker,

> "O'erlabored with his being's strife,
> Sink to that sweet forgetfulness of life!"

Neither the Colonel nor the man Jack moved an inch after taking their places. I almost envied them their capacity to sleep, so gentle and profound was their oblivion to the world and all its cares. To me this refreshing luxury was denied. My fate seemed to hang upon a thread. I could not feel any confidence in these men. They might become suspicious at any moment, and murder me as I lay helpless before them. For over two hours I watched them; they never moved. The probable fact was, they had made up their minds not to molest me, in view of the large sum of money I expected to collect at San Luis. My course seemed clear enough. But here was the difficulty. I could do nothing without my papers. Nor was I content to lose my mule, saddle, and blankets, which I knew to be in their possession.

The tall man, Griff, was restless, and turned repeatedly, moaning in his sleep, "God have pity on me! Oh God, have pity on me!"

It was a sad sight to behold him. No mortal eye could fathom the sufferings that thus moved him. Truly,

> "The mind that broods o'er guilty woes
> Is like a scorpion girt by fire."

At length—it must have been about an hour before day—he arose, looked cautiously around, and, seeing all quiet, beckoned to me, and stealthily left the camp. On his way out he gathered up my blanket, saddle, and coat in his arms, and looked back to see if I had taken the hint. I lost no time in slipping from my covering, and following his receding figure. It was a trying moment.

I expected to see the other two men rise, and held my pistol ready for defense. In a few minutes we were beyond immediate danger of discovery.

"Now," said Griff—"now is your time. Here is your mule. Mount him and be off! They will undertake to pursue you as soon as they discover your absence; but I shall loose the riatas, and it will take them some time to catch the horses. You will find your papers on the trail as soon as you strike the plain. Get to San Miguel, and you are safe. They dare not go there; *but don't stop on the way.*"

While he was talking Griff fixed my saddle and pack on the mule, and I mounted without loss of time. What could I do to reward this noble fellow? In the hurry of the moment I handed him my watch.

"Friend," said I, "you have done me an inestimable service. Take this trifle as a keepsake, and with it my best thanks. You and I may never meet again."

"No, it is not likely we shall," said Griff, sadly. "Our ways are different. Keep your watch; I can't accept it. All I ask of you is not to judge me harshly. Good-by!"

The impulse to serve this unfortunate man was irresistible. I could not leave him thus. It was no idle curiosity that prompted me to probe the mystery of his conduct.

"In heaven's name, friend, why do you stay with these bad men? What unholy power have they over you? Leave them, I implore you—leave them at once and forever. Come with me. I will do all I can for you. Surely you are not too far gone in crime for repentance. The vilest sinner may be saved!"

The poor fellow's frame was convulsed with agony. He sobbed like a child, and for a moment seemed unable to speak. Suddenly, as if recollecting himself, he said,

"No, sir, I can not turn traitor. It is no use—I am gone beyond redemption. Their fate must be mine. God pity me! I struggled hard against the evil spirit, but he has conquered. I am gone, sir—gone! Yet, believe me, I am not wholly depraved—a criminal in the eyes of the law; a robber; an outcast from society and civilization; but (here he lowered his voice to a whisper) —but NOT A MURDERER. Oh God, pity me! My mother —my poor old mother!"

This was all. The next moment he turned away, and was lost in the gloom of the trees.

CHAPTER VI.

A LONELY RIDE.

As I struck into the trail and out into the broad valley of the Salinas a sense of freedom relieved me in some degree of the gloom inspired by the last words of this strangely unfortunate man. The stars were shining brightly overhead, but the moon had gone down some time previously. It was just light enough to see the way. A small white object lying in the trail caused the mule to start. In the excitement of my escape I had forgotten about the papers. Here they were, all safe. I had no doubt they had been thus disposed of by the ruffian Jack during the previous evening when he took occasion to absent himself from the camp. I quickly dismounted and placed the package securely in the leg of one of my boots, then pushed on with all speed to reach a turning-point of the mountains some distance ahead, in order to be out of sight by the dawn of day, which could not be far off. In about an hour I had gained this point, and at the same time the first faint streaks of the coming day began to appear in the eastern sky. The air was peculiarly balmy—cool enough to be pleasant, and deliciously odorous with the herbage of the mountains. Already the deer began to leave their coverts among the shrubbery on the hill-sides, and numerous bands of them stood gazing at me as I passed, their antlers erect, their beautiful forms motionless, as if hewn

A LONELY RIDE.

from the solid rock, but manifesting more curiosity than fear. Thousands of rabbits frisked about in the open glades, and innumerable flocks of quail flitted from bush to bush. The field-larks and doves made the air musical with their joyous hymns of praise to the rising sun; the busy hum of bees rose among the wild flowers by the wayside; all nature seemed to awake from its repose smiling with a celestial joy. In no other country upon earth have I seen such mornings as in the interior of California—so clear, bright, and sparkling—so rich and glowing in atmospheric tints—so teeming with unbounded opulence in all that gives vigor, health, and beauty to animated nature, and inspiration to the higher faculties of man. There is a redundancy of richness in the earth, air, and light unknown even in that land of fascination which is said to possess "the fatal gift of beauty."

Contrasted with the dark spirit of crime that hung over my late encampment, such a morning was inexpressibly lovely. Every breath of air—every sound that broke upon the listening ear—every thought of the vast wild plains and towering mountains that swept around me in the immeasurable distance, inspired vague and unutterable sensations of pleasure and pain—pleasure that I was free and capable of enjoying such exquisite physical and mental luxuries; pain that here, on God's own footstool,

"All but the spirit of man was divine."

As the sun rose, and spread over mountain and valley a drapery of glowing light, giving promise of continued life to the birds of the air and the beasts of the field, I could not but think with sadness how man—made after God's own image, the most perfect of his works, gifted with reason and intelligence—should so strangely turn aside from the teachings of his Maker, and cast away the pure enjoyments so bountifully spread before him. Was it possible that a single created being, however steeped in crime, could be insensible to the soothing and humanizing influences of such a scene?

The unhappy fate of the poor fellow to whom I was

so deeply indebted haunted me. He, at least, must have felt the better promptings of his inner nature amid these beautiful works of a beneficent Creator. Surely such a man could never be utterly lost. There were noble traits in his character that must, some time or other, assert their supremacy. Honorable even in his degradation, he scorned to turn traitor to men whom he despised. His was not a nature formed for cruel and crafty deeds. Frank, manly, and ingenuous in his whole bearing, there was evidence of innate nobility in his misguided sense of honor, and a manifest scorn of deception in his wild outbursts of passion. What could have driven him to this career of crime? What satanic power was that by which he was enthralled? I could not believe that he was voluntarily bad. That single outburst of emotion as he spoke of his mother would have redeemed him had he been the worst of criminals. A career of dissipation must have brought him to this. He was evidently compromised, but to what extent? Some painful mystery hung over his connection with these bad men—I could not fathom it. The more I reflected upon all I had seen and heard, the more profound became my sympathy; nor is it an affectation of generosity to say that I would have sacrificed much to have saved him. Yet this man's case was not an uncommon one in California. There were many there, even at that early period, and there are still many, who, with the noblest attributes that adorn human nature, have become castaways.

As the day advanced a marked change became perceptible in the character of the country. Passing out from the valley of the Salinas to the right, the trail entered a series of smaller valleys, winding from one to another through a succession of narrow cañons between low, gravelly hills, destitute of shrubbery, and of a peculiarly whitish and barren aspect. The scene was no longer enlivened by bands of deer and smaller game, such as I had seen in the morning; the birds had also disappeared; not a living thing was in sight save a few buzzards

hovering in the air over the bleached ar
and occasionally a coyote or wild-cat skul
across the trail. Toward noon the earth be
fiery furnace. The air was scorching. In t
passages, where the hills converged into a foc
off every current of air, the refraction of the
was absolutely terrific. It seemed as if my
must crisp into tinder and drop from my body
skin peeled from my face and hands; a thick woole
was insufficient to keep the fierce and seething heat fr
my head, and I sometimes feared I would be smitten
the earth. Not knowing the water-holes, or rather hav
ing no time to look for them, I was parched with an intolerable thirst. On every eminence I turned to look back, but nothing was in sight save the dreary waste of barren hills that lay behind.

Toward evening, having stopped only a few minutes at a pool of water, my mule began to lag again. I had no spurs, and it was utterly in vain that I urged him on by kicks and blows. His greatest speed was a slow trot, and to keep that up for a few hundred yards at a time required my utmost efforts. By sundown I estimated that the distance to San Miguel must be twelve or fifteen miles. It was a very unpleasant position to be in—pursued, as I had every reason to suppose, by men who would not hesitate to take my life, yet unable to accelerate the speed of my animal. All I could do was to continue beating him.

The country became still more lonesome and desolate as I advanced. The chances of being overtaken momentarily increased. My anxiety to reach San Miguel caused me to forget all the sufferings of fatigue and thirst, and strain every nerve to get my mule over the ground. But the greater the effort the slower he traveled. It was true, I had a pistol, and could make some defense. Yet the chances were greatly against me. Unskilled in this sort of warfare, an indifferent rider, unacquainted with the trails by which I might be cut off and surprised, it seemed indeed a very hopeless case, should such an emergency arise. Besides, it would be very little satisfaction to shoot one, or even two men, against whom I felt no enmity, and whose lives were worth nothing to me, and still less to get killed myself. The truth is, I had a particular relish for life; others were interested in it as well as myself, and I did not feel disposed to risk it unnecessarily.

The sun went down at last, and the soft shadows of night began to soften the asperities of the scene. I rode on, never once relaxing my efforts to get a little more speed out of my mule. The moon rose, and innumerable stars twinkled in the sky. The air became delightfully balmy. Long shadows of rocks and trees swept across the trail. Mystic forms seemed to flit through the dim distance, or stand like ghostly sentinels along the way-side. Often I fancied I could see men on horseback stationed under the overhanging rocks, and detect the glitter of their arms in the moonlight. Stumps of trees riven by the storms of winter loomed up among the rocks like grim spectres; the very bushes assumed fantastic forms, and waved their long arms in gestures of warning. The howling of innumerable coyotes and the hooting of the night-owls had a singularly weird effect in the stillness of the night.

CHAPTER VII.

THE ATTACK.

It must have been nearly ten o'clock when my mule suddenly stopped, turned around, and set up that peculiar nickering bray by which these animals hail the approach of strangers. As soon as he ceased his unwelcome noise I listened, and distinctly heard the clatter of hoofs in the road, about half a mile in the rear. That my pursuers were rapidly approaching there was now very little doubt. It was useless to attempt to reach San Miguel, which must be still four or five miles distant. I had no time, and resolved at once to make for a little grove some three or four hundred yards to the right. As I approached the nearest trees I was rejoiced to see something like a fence. A little farther on was a gray object with a distinct outline. It must be a house. There was no light; but I soon discovered that I was within fifty yards of a small adobe building. My mule now pricked up his ears, snuffed the air wildly, and absolutely refused to move a step nearer. I dismounted, and tried to drag him toward the door. His terror seemed unconquerable. With starting eyes, and a wild blowing sound from his nostrils, he broke away and dashed out into the plain. I speedily lost sight of him.

This time I had taken the precaution to secure my papers and pistol on my person. The mule had taken the direction of San Miguel; but, even should I be unable to recover him, the loss would not be so great as before. However, it was no time to calculate losses. The clatter of hoofs grew nearer and nearer, and soon the advancing forms of two mounted men became distinctly visible in the moonlight. There was no alternative but to seek security in the old adobe. I ran for the door and pushed it open. The house was evidently untenanted. No answer was made to my summons save a mocking echo from the bare walls. My pursuers must have caught sight of me as they approached. I could hear their imprecations as they tried to force their animals up to the door. One of the party—the Colonel, whose voice I had no difficulty in recognizing, said,

"Blast the fellow! what did he come here for?"

The other answered with an oath and a brutal laugh,

"We've got him holed, any how. It won't take long to root him out."

They then dismounted and proceeded to tie their horses to the nearest tree. I could hear them talk as they receded, but could not make out what they said.

While this was going on I had closed the door, and was looking for some bolt or fastening, when I heard the low, fierce growl of some animal. There was no time to conjecture what it was; the next moment a furry skin brushed past, and the animal sprang through an opening in the wall.

A wooden bar was all I could find; but the iron fastening had been broken, and the only way of securing the door was to brace the bar against it in a diagonal position. The floor was of rough hard clay, and served in some sort to prevent the brace from slipping. A few moments of painful anxiety passed. I had drawn my revolver, and stood close against the inner wall, prepared to fire upon the first man that entered. Presently the two men returned, approaching stealthily along the wall, so as to avoid coming in range of the door. The sharp, hard voice of the Colonel first broke the silence.

"Come," said he, "open the door! You can't help yourself now! It is all up with you, my fine fellow!"

I knew the villains wanted to find my position, and made no answer.

"You may as well come out at once," said the Colonel; "you have no chance. There is nobody here to stand by you as there was last night. Your friend is keeping camp with a bullet through his head and a gash in his throat."

Pressed as I was, this news shocked me beyond measure. The unfortunate man who had befriended me had paid the penalty of his life for his kindness.

"Out with you!" roared the Colonel, fiercely, "or we'll burst the door down. Come, be quick!"

Another pause. I heard a low whispering, and stood with breathless anxiety with my finger upon the trigger of my pistol. In that brief period it was wonderful how many thoughts flashed through my mind. I knew nothing of the construction of the house; had no time even to look around and see if there was any back entrance. A faint light through one small window-hole in front,

THE ATTACK.

K

within three feet of the door, was all I could discern. Every nerve was strained to its utmost tension. My sense of hearing was painfully acute. The low whispering of the two ruffians, the faint jingling of their spurs, the very creaking of their boots, as they stealthily moved, was fearfully audible. With an almost absolute certainty of death, without the remotest hope of relief, it was strange how my thoughts wandered back upon the past; how the peaceful fireside of home was pictured to my mind; how vividly I saw the beloved faces of kindred and friends; how all that were dear to me seemed to sympathize in my unhappy fate. Yet it was impossible to realize that my time had come. The whole thing—the camp, the dark, murderous faces, the chase, the blockade—resembled rather some horrible fantasy than the dread truth. Strange, too, that I should have noticed something even grotesque in my situation; run into a hole, as the ruffian Jack had said, like a coyote or a badger. Five minutes—it seemed a long time—must have passed in this way, when I became conscious of a gradual darkening in the room. A low, heavy breathing attracted my attention. I looked in the direction of the window, and thought I could detect something moving; but the darkness was so impenetrable that it might be the result of imagination. Should I fire and miss my mark, the flash would reveal my position and be certain destruction. The dark mass again moved. I could distinctly hear the respiration. It must be one of the men trying to get in through the small window-hole. I raised my pistol, took dead aim as near as possible upon the centre of the object, and fired. The fall of a heavy body outside, a groan, an imprecation, was all I could hear, when a tremendous effort was made to force the door, and two shots were fired through it in quick succession. The wood was massive, but much decayed; and I saw that it was rapidly giving way before the furious assaults that were made upon it from the outside, evidently with a heavy piece of timber. Another lunge or two of this powerful battering-ram must have borne it from its hinges or shattered it to fragments.

"Hold on, Jack!" said the wounded man in a low voice; "come here, quick! The infernal fool has shot me through the shoulder! I'm bleeding badly."

The ruffian dropped his bar, as I judged by the sound, and turned to drag his leader out of range of the door. Now was the time for a bold move. Hitherto I had acted on the defensive; but every thing depended on following up the advantage. Removing the brace from the door, I made an opening sufficient to get a glimpse of the two men. The stout fellow, Jack, was stooping down, dragging the other toward the corner of the house. I fired again. The ball was too low; it missed his body, but must have shattered his wrist; for, with a horrible oath, he dropped his burden, and staggered back a few paces writhing with pain, his hand covered with blood. Before I could get another shot he darted behind the house. At the same time the Colonel rose on his knee, turned quickly, and fired. The ball whizzed by my head and struck the door. While I was trying to get a shot at him in return, he jumped to his feet and staggered out of range. I thought it best now to rest satisfied with my success so far, and again retired to my position behind the door.

For the next ten or fifteen minutes I could hear, from time to time, the smothered imprecations of the wounded ruffians, but after this there was a dead silence. I heard nothing more. They had either gone or were lying in wait near by, supposing I would come out. This uncertainty caused me considerable anxiety, for I dared not abandon my gloomy retreat. Two or three hours must have passed in this way, during which I was constantly on the guard; but not the slightest indication of the presence of the enemy was perceptible.

Two nights had nearly passed, during which I had not closed my eyes in sleep. The perpetual strain of mind and the fatigue of travel were beginning to tell. I felt faint and drowsy. During the whole terrible ordeal of this night I had not dared to sit down. But now my legs refused to support me any longer. I groped my way toward a corner of the room to lie down. Some soft mass on the ground caused me to stumble. I threw out my hands and fell. What was it that sent such a thrill of horror through every fibre? A dead body lay in my embrace—cold, mutilated, and clotted with blood!

It has been my fortune, during a long career of travel in foreign lands, to see death in many forms. I do not profess to be exempt from the weakness common to most men—a natural dread of that undiscovered region toward which we are all traveling. But I never had any peculiar repugnance to the presence of dead men. What are they, after all, but inanimate clay? The living are to be feared—not the dead, who sleep the sleep that knows no waking. Not this—not the sudden contact with a corpse; not simply the cold and blood-clotted face over which I passed my hand was it that caused me to recoil with such a thrill of horror. It was the solution of a dread mystery. There, in a pool of clotted gore, lay the corpse of a murdered man. No need was there to conjecture who were his murderers.

I rose up, thoroughly aroused from my drowsiness. It was probable others had shared the fate of this man: If so, their bodies must be near at hand. I was afraid to open the door to let in the light, for, bad as it was to be shut up in a dark room with the victim or victims of a cruel murder, it was worse to incur the risk of a similar fate by exposing myself. After somewhat recovering my composure I groped about, and soon discovered that three other bodies were lying in the room: one on a bed—a woman with her throat cut from ear to ear—and two smaller bodies on the floor near by—children perhaps eight or ten years old, but so mutilated that it was difficult to tell what they were. Their limbs were almost denuded of flesh, and their faces and bodies were torn into shapeless masses. This must have been the finishing work of the animal—a coyote no doubt—that had startled me with a growl, and broken through the window after I had first closed the door. I could also now account for the strange manner in which the mule had snuffed the air, and his unconquerable terror in approaching the house.

Only a few articles of furniture were in the room—a bed, two or three broken stools, a frying-pan, coffee-pot, and a few other cooking utensils, thrown in a heap near the fireplace. There was no other room; nor was there any back entrance, as I had at first apprehended.

It was a gloomy place enough to spend a night in, but there was no help for it. I certainly had less fear of the dead than of the living. It could not be over two or three hours till morning; and it was not likely the two men, who were seeking my life, would lurk about the premises much longer, if they had not long since taken their departure, which seemed the most probable.

I knelt down and commended my soul to God; then stretched myself across the brace against the door, and, despite the presence of death, fell fast asleep. It was broad daylight when I awoke. The sun's earliest rays were pouring into the room through the little window and the cracks of the door. A ghastly spectacle was revealed—a ghastly array of room-mates lying stiff and stark before me.

From the general appearance of the dead bodies I judged them to be an emigrant family from some of the Western States. They had probably taken up a temporary residence in the old adobe hut after crossing the plains by the southern route, and must have had money or property of some kind to have inspired the cupidity of their murderers. The man was apparently fifty years of age; his skull was split completely open, and his brains scattered out upon the earthen floor. The woman was doubtless his wife. Her clothes were torn partly from her body, and her head was cut nearly off from her shoulders; besides which, her skull was fractured with some dull instrument, and several ghastly wounds disfigured her person. The bedclothes were saturated with blood, now clotted by the parching heat. The two children had evidently been cut down by the blows of an axe. Their heads were literally shattered to fragments. What the murderers had failed to accomplish in mutilating the bodies had been completed by some ravenous beast of prey—the same, no doubt, already mentioned.

I saw no occasion to prolong my stay. It was hardly probable the Colonel and Jack, wounded as they were, would renew their attack. They must have made their way back to camp, or at least retired to some part of the country where they would incur less risk of capture.

CHAPTER VIII.

SAN MIGUEL.

It was a bright and beautiful morning as I left the house and turned toward San Miguel. The contrast between the peaceful scene before me and the horrible sight I had just witnessed was exceedingly impressive. The mellow light of the early sun on the mountains; the winding streams fringed with shrubbery; the rich, golden hue of the valley; the cattle grazing quietly in the low meadows bordering on the Salinas River; the singing of the birds in the oak groves, were indescribably refreshing to a fevered mind, and filled my heart with thankfulness that I was spared to enjoy them once more. Yet I could not but think of what I had witnessed in the adobe hut—a whole family cut down by the ruthless hands of murderers who might still be lurking behind the bushes on the wayside. Their dreadful crime haunted the scene, and its exquisite repose seemed almost a cruel mockery. De Quincey somewhere remarks that he never experienced such profound sensations of sadness as on a bright summer day, when the very luxuriance and maturity of outer life, and the fullness of sunshine that filled the visible world, made the desolation and the darkness within the more oppressive. I could now well understand the feeling; and though grief had but little part in it, beyond a natural regret for the unhappy fate of the murdered family, still it was sad to feel the contrast between the purity and beauty of God's creation and the willful wickedness of man.

I had not lost the strong instinct of self-preservation, which, so far at least, through the kind aid of Providence, had enabled me to preserve my life; and in my lonely walk toward San Miguel I was careful to keep in the open valley, and avoid, as much as possible, coming within range of the rocks and bushes. In about an hour I saw the red tile roofs and motley collection of ruinous old buildings that comprised the former missionary station of San Miguel. A gang of lean wolfish dogs ran out to meet me as I approached, and it was not without difficulty that I could keep them off without resorting to my revolver, which was an alternative that might produce a bad impression where I most hoped to meet with a friendly reception. As I approached the main buildings I was struck with the singularly wild and desolate aspect of the place. Not a living being was in sight. The carcass of a dead ox lay in front of the door, upon which a voracious brood of buzzards were feeding; and a coyote sat howling on an eminence a little beyond. I walked into a dark, dirty room, and called out, in what little Spanish I knew, for the man of the house. "*Quien es?*" demanded a gruff voice. I looked in a corner, and saw a filthy-looking object, wrapped in a poncho, sitting lazily on a bed. By his uncouth manner and forbidding appearance I judged him to be the vaquero in charge of the place, in which I was not mistaken. With considerable difficulty I made him comprehend that I had lost my mule, and supposed it had strayed to San Miguel.

"*Quien sabe?*" said the fellow, indifferently.

Could he not find it? I would be willing to reward

SAN MIGUEL.

him. I would give him the blankets. I was an *Oficiál*, and was on my way to San Luis Obispo. To each of these propositions the man returned a stupid and yawning answer, "*Quien sabe*—who knows?"

Finding nothing to be gained on that point, I asked him for something to eat, for I was well-nigh famished with hunger. He pointed lazily to a string of jerked beef strung across the rafters. It required but little time to select a few dry pieces, and while I was eating them the fellow asked me if I had any tobacco. I handed him a plug, which speedily produced a good effect, for he got up and passed me a plate of cold tortillas. When I had somewhat satisfied the cravings of hunger, I asked him, in my broken Spanish, if he had heard of the murder—five persons killed in an old adobe house near by. "*Quien sabe?*" said he, in the same indifferent tone. "*Muchos malhos hombres aqui.*" This was all he knew, or professed to know, of the murder.

"Amigo," said I, "if you'll get my mule and bring him here, I'll give you this watch."

He took the watch and examined it carefully, handed it back, and remarked as before, "*Quien sabe?*" The glitter of the gold, however, seemed to quicken his perceptive faculties to this extent that he got up from the bed, put on his spurs, took a riata from a peg on the wall, and walked out, leaving me to entertain myself as I thought proper during his absence.

Having finished a substantial repast of jerked beef and tortillas, I went out and rambled about among the ruins for nearly an hour. A few lazy and thriftless Indians, lying in the sun here and there, were all the inhabitants of the place I could see. This ranch must have been a very desirable residence in former times. The climate is charming, except that it was a little warm in summer, and the cattle ranges are richly clothed with grass and very extensive.

In about an hour my friend the vaquero came back, mounted on a broncho or wild horse, leading after him

A SPANISH CABALLERO.

my mule, with the pack unchanged. From what I could understand, he had found the mule entangled by the bridle in the bushes, some three miles on the trail toward San Luis. According to promise, I handed him my watch. He took it and examined it again, then handed it back without saying a word.

"*Amigo*," said I, "the watch is yours. I promised it to you if you found my mule."

To this he merely shrugged his shoulders.

"Won't you take it? I have no money."

"No, señor," said he, at length, with a somewhat haughty air, "I am a Spanish gentleman."

"Oh, I beg your pardon. Will you do me the favor, then, to accept a plug of tobacco?"

I opened my pack and handed him a large plug of the finest pressed Cavendish.

"*Mil gracias!*" said the Spanish gentleman, smiling affably, and making a condescending inclination of the head. "That suits me better. A watch is bad property here. I don't want to be killed yet a while."

Here was a hint of his reason for declining the proffered reward. But he did it very grandly; and I was quite willing to accord to him the title of Señor Caballero to which he aspired, though he certainly looked as unlike the Caballeros described by the learned Fray Antonio Agapida, who went out to make war upon the Moors of Granada, as one distinguished individual can look unlike another.

There was ample reason why I should regard my mule with dissatisfaction. All my misfortunes, so far, had arisen from his defective physical and mental organization (if I may use the term in reference to such an animal); but the fact is, it has been my fate, as far back as I can recollect, to have the worst stock in the country foisted upon me. Never yet, up to this hour, have I succeeded in purchasing a sound, safe, and reliable animal—except, indeed, an old horse that I once owned in Oakland, generally known in the neighborhood as Selim the Steady

—a name derived from his unconquerable propensity for remaining in the stable, or getting back to it as soon as ever he left the premises.

The vaquero, or, as he aspired to be called, the Caballero, offered to barter his broncho for my mule, and, as an inducement, set him to bucking all over the ground within a circle of fifty yards, merely to show the spirit of the animal, of which I was so well satisfied that I declined the barter.

CHAPTER IX.

A DANGEROUS ADVENTURE.

Bidding my worthy friend a kindly "adios," I mounted the mule and pursued my journey toward San Luis. The country, for many miles after leaving San Miguel, was very wild and picturesque. Blue mountains loomed up in the distance; and the trail passed through a series of beautifully undulating valleys, sometimes extensive and open, but often narrowed down to a mere gorge between the irregular spurs of the mountains. Game was very abundant, especially quail and rabbits. I saw also several fine herds of deer, and occasionally bands of large red wolves. It was a very lonesome road all the way to the valley of Santa Marguerita, not a house or human being to be seen for twenty miles at a stretch. Toward evening, on the first day after leaving San Miguel, I descended the bed of a creek to water my mule. While looking for the water-hole, I heard some voices, and suddenly found myself close by a camp of Sonoranians. It was too late to retreat, for I was already betrayed by the braying of my mule. Upon riding into the camp I was struck with the savage and picturesque group before me, consisting of some ten or a dozen Sonoranians. It is doing them no more than justice to say that they were the most villainous, cut-throat, ill-favored looking gang of vagabonds I had ever laid eyes upon. Some were smoking cigarritos by the fire, others lying all about the trees playing cards, on their ragged saddle-blankets, with little piles of silver before them; and those that were not thus occupied were capering around on wild horses, breaking them apparently, for the blood streamed from the nostrils and flanks of the unfortunate animals, and they were covered with a reeking sweat.

Probably it may be thought that I exceeded the truth when I asked this promising party if they had seen six "Americanos" pass that way with a pack-train from San Luis, friends of mine that I was on the look-out for. They had seen no such pack-train; it had not passed since they camped there, which was several days ago.

"Then," said I, "it must be close at hand, and I must hurry on to meet it. The mules are laden with *mucha plata.*"

Having watered my mule, I rode on about five miles farther, where I reached a small ranch-house occupied by a native Californian family. They gave me a good supper of frijoles and jerked beef, and I slept comfortably on the porch.

Next day I struck into the Valley of Santa Marguerita. I shall never forget my first impression of this valley. Encircled by ranges of blue mountains were broad, rich pastures, covered with innumerable herds of cattle; beautifully diversified with groves, streams, and shrubbery; castellated cliffs in the foreground as the trail wound downward; a group of cattle grazing by the margin of a little lake, their forms mirrored in the water; a mirage in the distance; mountain upon mountain beyond, as far as the eye could reach, till their dim outlines were lost in the golden glow of the atmosphere. Surely a more lovely spot never existed upon earth. I have wandered over many a bright and beautiful land, but never, even in the glorious East, in Italy, Spain, Switzerland, or South America, have I seen a country so richly favored by nature as California, and never a more lovely valley than

VALLEY OF SANTA MARGUERITA.

Santa Marguerita upon the whole wide world. There is nothing comparable to the mingled wildness and repose of such a scene; the rich and glowing sky, the illimitable distances, the teeming luxuriance of vegetation, its utter isolation from the busy world, and the dreamy fascination that lurks in every feature.

I had passed nearly across the valley, and was about to enter upon an undulating and beautifully timbered range of country extending into it from the foot-hills, when a dust arose on a rise of ground a little to the left and about half a mile distant. My mule, ever on the alert for some new danger, pricked up his ears and manifested symptoms of uncontrollable fear. The object rapidly approached, and without farther warning the mule whirled around and fled at the top of his speed. Neither bridle nor switch had the slightest effect. In vain I struggled to arrest his progress, believing this, like many other frights he had experienced on the road, was rather the result of innate cowardice than of any substantial cause of apprehension. One material difference was perceptible. He never before ran so fast. Through brush and mire, over rocks, into deep arroyas and out again, he dashed in his frantic career, never once stopping till by some mischance one of his fore feet sank in a squirrel-hole, when he rolled headlong on the ground, throwing me with considerable violence several yards in advance. I jumped to my feet at once, hoping to catch him before he could get up, but he was on his feet and away before I had time to make the attempt. It now became a matter of personal interest to know what he was running from. Upon looking back, I was astonished to see not only one object, but four others in the rear, bearing rapidly down toward me. The first was a large animal of some kind — I could not determine what — the others mounted horsemen in full chase. Whatever the object of the chase was, it was not safe to be a spectator in the direct line of their route. I cast a hurried look around, and discovered a break in the earth a few hundred yards distant, toward which I ran with all speed. It was a sort of mound rooted up by the squirrels or coyotes, and afforded some trifling shelter, where I crouched down close to the ground. Scarcely had I partially concealed myself when I heard a loud shouting from the men on horseback, and, peeping over the bank, saw within fifty or sixty paces a huge grizzly bear, but no longer retreating. He had faced round toward his pursuers, and now seemed determined to fight. The horsemen were evidently native Californians, and managed their animals with wonderful skill and grace. The nearest swept down like an avalanche toward the bear, while the others coursed off a short distance in a circling direction to prevent his escape. Suddenly swerving a little to one side, the leader whirled his lasso once or twice around his head, and let fly at his game with unerring aim. The loop caught one of the fore paws, and the bear was instantly jerked down upon his haunches, struggling and roaring with all his might. It was a striking instance of the power of the rider over the horse, that, wild with terror as the latter was, he dared not disobey the slightest pressure of the rein, but went through all the evolutions, blowing trumpet-blasts from his nostrils and with eyes starting from their sockets. Despite the strain kept upon the lasso, the bear soon regained his feet, and commenced hauling in the spare line with his fore paws so as to get within reach of the horse. He had advanced within ten feet before the nearest of the other horsemen could bring his lasso to bear upon him. The first throw was at his hind legs—the main object being to stretch him out—but it missed. Another more fortunate cast took him round the neck. Both riders pulled in opposite directions, and the bear soon rolled on the ground again, biting furiously at the lassos, and uttering the most terrific roars. The strain upon his neck soon choked off his breath, and he was forced to let loose his grasp upon the other lasso. While struggling to free his neck, the two other horsemen dashed up, swinging their lassos, and

LASSOING A GRIZZLY.

shouting with all their might so as to attract his attention. The nearest, watching narrowly every motion of the frantic animal, soon let fly his lasso, and made a lucky hitch around one of his hind legs. The other, following quickly with a large loop, swung it entirely over the bear's body, and all four riders now set up a yell of triumph and began pulling in opposite directions. The writhing, pitching, and straining of the powerful monster were now absolutely fearful. A dust arose over him, and the earth flew up in every direction. Sometimes by a desperate effort he regained his feet, and actually dragged one or more of the horses toward him by main strength; but, whenever he attempted this, the others stretched their lassos, and either choked him or jerked him down upon his haunches. It was apparent that his wind was giving out, partly by reason of the long chase, and partly owing to the noose around his throat. A general pull threw him once more upon his back. Before he could regain his feet, the horsemen, by a series of dexterous manœuvres, wound him completely up, so that he lay perfectly quiet upon the ground, breathing heavily, and utterly unable to extricate his paws from the labyrinth of lassos in which he was entangled. One of the riders now gave the reins of his horse to another and dismounted. Cautiously approaching, with a spare riata, he cast a noose over the bear's fore paws, and wound the remaining part tightly round the neck, so that what strength might still have been left was speedily exhausted by suffocation. This done, another rider dismounted, and the two soon succeeded in binding their victim so firmly by the paws that it was impossible for him to break loose. They next bound his jaws together by means of another riata, winding it all the way up around his head, upon which they loosened the fastening around his neck so as to give him air. When all was secure, they freed the lassos and again mounted their horses. I thought it about time now to make known my presence and stood up. Some of the party had evidently seen me during the progress of the chase, for they manifested no surprise; and the leader, after exchanging a few words with one of the men, and pointing in the direction taken by the mule, rode up and said very politely,

"*Buenas dias, Señor!*" He then informed me, as well as I could understand, that he had sent a man to catch my mule, and it would be back presently. While we were endeavoring to carry on some conversation in reference to the capture of the bear, during which I made out to gather that they were going to drag him to the ranch on a bullock's hide, and have a grand bull-fight with him in the course of a few days, the vaquero returned with my mule.

I had a pleasant journey of thirty-five miles that day. Nothing farther occurred worthy of record. When night overtook me I was within fifteen miles of San Luis. I camped under a tree, and, notwithstanding some apprehension of the Sonoranians, made out to get a good sleep.

Next morning I was up and on my way by daylight. The country, as I advanced, increased in picturesque beauty, and the hope of soon reaching my destination gave me additional pleasure. A few hours more, and I was safely lodged with some American friends. Thus ended what I think the reader must admit was "a dangerous journey."

CHAPTER X.

A TRAGEDY.

A FEW days after my arrival in San Luis I went, in company with a young American by the name of Jackson, to a fandango given by the native Californians. The invitation, as usual in such cases, was general, and the company not very select. Every person within a circle of twenty miles, and with money enough in his pockets to pay for the refreshments, was expected to be present. The entertainment was held in a large adobe building, formerly used for missionary purposes, the lower part of which was occupied as a store-house. A large loft overhead, with a step-ladder reaching to it from the outside, formed what the proprietor was pleased to call the dancing-saloon. In the yard, which was encircled by a mud wall, were several chapadens, or brush tents, in which whisky, gin, aguardiente, and other refreshments of a like nature, "for ladies and gentlemen," were for sale at "two bits a drink." A low rabble of Mexican greasers, chiefly Sonoranians, hung around the premises in every direction, among whom I recognized several belonging to the gang into whose encampment I had fallen on my way down from Santa Marguerita. Their dirty serapas, machillas, and spurs lay scattered about, just as they had dismounted from their mustangs. The animals were picketed around in the open spaces, and kept up a continual confusion by bucking and kicking at every straggler who came within their reach. Such of the rabble as were able to pay the entrance-fee of "*dos realles*" were sitting in groups in the yard, smoking cigarritos and playing at monté. A few of the better class of rancheros had brought señoritas with them, mounted in front on their saddles, and were wending their way up the step-ladder as we entered the premises.

I followed the crowd, in company with my friend Jackson, and was admitted into the saloon upon the payment of half a dollar. This fund was to defray the expense of lights and music.

On passing through the doorway I was forcibly impressed with the scene. Some fifty or sixty couples were dancing to the most horrible scraping of fiddles I had ever heard, marking the time by snapping their fingers, whistling, and clapping their hands. The fiddles were accompanied by a dreadful twanging of guitars; and an Indian in one corner of the saloon added to the din by beating with all his might upon a rude drum. There was an odor of steaming flesh, cigarritos, garlic, and Cologne in the hot, reeking atmosphere that was almost suffocating; and the floor swayed under the heavy tramp of the dancers, as if every turn of the waltz might be the last. The assemblage was of a very mixed character, as may well be supposed, consisting of native Californians, Sonoranians, Americans, Frenchmen, Germans, and half-breed Indians.

Most of the Mexicans were rancheros and vaqueros from the neighboring ranches, dressed in the genuine style of Caballeros del Campaña, with black or green velvet jackets, richly embroidered; wide pantaloons, open at the sides, ornamented with rows of silver buttons; a red sash around the waist; and a great profusion of gold filigree on their vests. These were the fast young fellows who had been successful in jockeying away their horses, or gambling at monté. Others of a darker and lower grade, such as the Sonoranians, wore their hats and machillas just as they had come in from camp; for it was one of the privileges of the fandango that every man could dress or undress as he pleased. A very desperate and ill-favored set these were—perfect specimens of Mexican outlaws.

The Americans were chiefly a party of Texans, who had recently crossed over through Chihuahua, and compared not unfavorably with the Sonoranians in point of savage costume and appearance. Some wore broadcloth frock-coats, ragged and defaced from the wear and tear of travel; some red flannel shirts, without any coats—their pantaloons thrust in their boots in a loose, swaggering style; and all with revolvers and bowie-knives swinging from their belts. A more reckless, devil-may-care looking set it would be impossible to find in a year's journey. Take them altogether—with their uncouth costumes, bearded faces, lean and brawny forms, fierce, savage eyes, and swaggering manners—they were a fit assemblage for a frolic or a fight. Every word they spoke was accompanied by an oath. The presence of the females imposed no restraint upon the subject or style of the conversation, which was disgusting to the last degree. I felt ashamed to think that habit should so brutalize a people of my own race and blood.

Many of the señoritas were pretty, and those who had no great pretensions to beauty in other respects were at least gifted with fine eyes and teeth, rich brunette complexions, and forms of wonderful pliancy and grace. All, or nearly all, were luminous with jewelry, and wore dresses of the most flashy colors, in which flowers, lace, and glittering tinsel combined to set off their dusky charms. I saw some among them who would not have compared unfavorably with the ladies of Cadiz, perhaps in more respects than one. They danced easily and naturally; and, considering the limited opportunity of culture they had enjoyed in this remote region, it was wonderful how free, simple, and graceful they were in their manners.

The belle of the occasion was a dark-eyed, fierce-looking woman of about six-and-twenty, a half-breed from Santa Barbara. Her features were far from comely, being sharp and uneven; her skin was scarred with fire or small-pox; and her form, though not destitute of a certain grace of style, was too lithe, wiry, and acrobatic to convey any idea of voluptuous attraction. Every motion, every nerve seemed the incarnation of a suppressed vigor; every glance of her fierce, flashing eyes was instinct with untamable passion. She was a mustang in human shape—one that I thought would kick or bite upon very slight provocation. In the matter of dress she was almost Oriental. The richest and most striking colors decorated her, and made a rare accord with her wild and singular physique; a gorgeous silk dress of bright orange, flounced up to the waist; a white bodice, with blood-red ribbons upon each shoulder; a green sash around the waist; an immense gold-cased breast-pin, with diamonds glittering in the centre, the greatest profusion of rings on her fingers, and her ears loaded down with sparkling

THE BELLE OF THE FANDANGO.

ear-rings; while her heavy black hair was gathered up in a knot behind, and pinned with a gold dagger—all being in strict keeping with her wild, dashing character, and bearing some remote affinity to a dangerous but royal game-bird. I thought of the Mexican chichilaca as I gazed at her. There was an intensity in the quick flash of her eye which produced a burning sensation wherever it fell. She cast a spell around her not unlike the fascination of a snake. The women shunned and feared her; the men absolutely worshiped at her shrine. Their infatuation was almost incredible. She seemed to have some supernatural capacity for arousing the fiercest passions of love, jealousy, and hatred. Of course there was great rivalry to engage the hand of such a belle for the dance. Crowds of admirers were constantly urging their claims. It was impossible to look upon their excited faces and savage rivalry, knowing the desperate character of the men, without a foreboding of evil.

"Perhaps you will not be surprised," said Jackson, "to hear something strange and startling about that woman. She is a murderess! Not long since she stabbed to death a rival of hers, another half-breed, who had attempted to win the affections of her paramour. But, worse than that—she is strongly suspected of having killed her own child a few months ago, in a fit of jealousy caused by the supposed infidelity of its father—whose identity, however, can not be fixed with any certainty. She is a strange, bad woman—a devil incarnate; yet you see what a spell she casts around her! Some of these men are mad in love with her! They will fight before the evening is over. Yet she is neither pretty nor amiable. I can not account for it. Let me introduce you."

As soon as a pause in the dance occurred I was introduced. The revolting history I had heard of this woman inspired me with a curiosity to know how such a fiend in human shape could exercise such a powerful sway over every man in the room.

Although she spoke but little English, there was a peculiar sweetness in every word she uttered. I thought I could detect something of the secret of her magical powers in her voice, which was the softest and most musical I had ever heard. There was a wild, sweet, almost unearthly cadence in it that vibrated upon the ear like the strains of an Æolian. Added to this, there was a power of alternate ferocity and tenderness in her deep, passionate eyes, that struck to the inner core wherever she fixed her gaze. I could not determine for my life which she resembled most—the untamed mustang, the royal game-bird, or the rattlesnake. There were flitting hints of each in her, and yet the comparison is feeble and inadequate. Sometimes she reminded me of Rachel—then the living, now the dead, Queen of Tragedy. Had it not been for a horror of her repulsive crimes, it is hard to say how far her fascinating powers might have affected me. As it was, I could only wonder whether she was most genius or devil. Not knowing how to dance, I could not offer my services in that way, and, after a few commonplace remarks, withdrew to a seat near the wall. The dance went on with great spirit. Absurd as it may seem, I could not keep my eyes off this woman. Whichever way she looked there was a commotion—a shrinking back among the women, or the symptoms of a jealous rage among the men. For her own sex she manifested an absolute scorn; for the other she had an inexhaustible fund of sweet glances, which each admirer might take to himself.

At a subsequent period of the evening I observed, for the first time, among the company a man of very conspicuous appearance, dressed in the very picturesque style of a Texan Ranger. His face was turned from me when I first saw him, but there was something manly and imposing about his figure and address that attracted my attention. While I was looking toward him he turned to speak to some person near him. My astonishment may well be conceived when I recognized in his strongly-marked features and dejected expression the face of the

L

man "Griff," to whom I was indebted for my escape from the assassins near Soledad! There could be no doubt that this was the outlaw who had rendered me such an inestimable service, differently dressed, indeed, and somewhat disfigured by a ghastly wound across the temple, but still the same; still bearing himself with an air of determination mingled with profound sadness. It was evident the Colonel had misinformed me as to his death. Perhaps, judging from the wound on his temple, which was still unhealed, he might have been left for dead, and subsequently have effected his escape. At all events, there was no doubt that he now stood before me.

I was about to spring forward and grasp him by the hand, when the dreadful scene I had witnessed in the little adobe hut near San Miguel flashed vividly upon my mind, and, for the moment, I felt like one who was paralyzed. That hand might be stained with the blood of the unfortunate emigrants! Who could tell? He had disavowed any participation in the act, but his complicity, either remote or direct, could scarcely be doubted from his own confession. How far his guilt might render him amenable to the laws I could not of course conjecture. It was enough for me, however, that he had saved my life; but I could not take his hand.

While reflecting upon the course that it might become my duty to pursue under the circumstances, I observed that he was not exempt from the fascinating sway of the dark señorita, whose face he regarded with an interest even more intense than that manifested by her other admirers. He was certainly a person calculated to make an impression upon such a woman; yet, strange to say, he was the only man in the crowd toward whom she evinced a spirit of hostility. Several times he went up to her and asked her to dance. Whether from caprice or some more potent cause I could not conjecture, but she invariably repulsed him—once with a degree of asperity that indicated something more than a casual acquaintance. It was in vain he attempted to cajole her.

She was evidently bitter and unrelenting in her animosity. At length, incensed at his pertinacity, she turned sharply upon him, and leaning her head close to his ear, whispered something, the effect of which was magical. He staggered back as if stunned, and, gazing a moment at her with an expression of horror, turned away and walked out of the room. The woman's face was a shade paler, but she quickly resumed her usual smile, and otherwise manifested no emotion.

This little incident was probably unnoticed by any except myself. I sat in a recess near the window, and could see all that was going on without attracting attention. I had resolved, after overcoming my first friendly impulses, not to discover myself to the outlaw until the fandango was over, and then determine upon my future course regarding him by the result of a confidential interview. I fully believed that he would tell me the truth, and nothing but the truth, in reference to the murder of the emigrants.

The dance went on. It was a Spanish waltz; the click-clack of the feet, in slow-measured time, was very monotonous, producing a peculiarly dreamy effect. I sometimes closed my eyes and fancied it was all a wild, strange dream. Visions of the beautiful country through which I had passed flitted before me—a country desecrated by the worst passions of human nature. Amid the rarest charms of scenery and climate, what a combination of dark and deadly sins oppressed the mind! What a cess-pool of wickedness was here within these very walls!

Half an hour may have elapsed in this sort of dreaming, when Griff, who had been so strangely repulsed by the dark señorita, came back and pushed his way through the crowd. This time I noticed that his face was flushed, and a gleam of desperation was in his eye. The wound in his temple had a purple hue, and looked as if it might burst out bleeding afresh. His motions were unsteady—he had evidently been drinking. Edging over toward the woman, he stood watching her till there was a pause in the dance. Her partner was a handsome young Mexican, very gayly dressed, whom I had before noticed, and to whom she now made herself peculiarly fascinating. She smiled when he spoke; laughed very musically at every thing he said; leaned up toward him, and assumed a wonderfully sweet and confidential manner. The Mexican was perfectly infatuated. He made the most passionate avowals, scarcely conscious what he was saying. I watched the tall Texan. The veins in his forehead were swollen; he strode to and fro restlessly, fixing fierce and deadly glances upon the loving couple. A terrible change had taken place in the expression of his features, which ordinarily had something sweet and sad in it. It was now dark, brutish, and malignant. Suddenly, as if by an ungovernable impulse, he rushed up close to where they stood, and, drawing a large bowie-knife, said to the woman, in a quick, savage tone,

"Dance with me now, or, damn you, I'll cut your heart out!"

She turned toward him haughtily—"Señor!"

"Dance with me, OR DIE!"

"Señor," said the woman, quietly, and with an unflinching eye, "you are drunk! Don't come so near to me!"

The infuriated man made a motion as if to strike at her with his knife; but, quick as lightning, the young Mexican grasped his uprisen arm and the two clenched. I could not see what was done in the struggle. Those of the crowd who were nearest rushed in, and the affray soon became general. Pistols and knives were drawn in every direction; but so sudden was the fight that nobody seemed to know where to aim or strike. In the midst of the confusion, a man jumped up on one of the benches and shouted,

"Back! back with you! The man's stabbed! Let him out!"

The swaying mass parted, and the tall Texan staggered through, then fell upon the floor. His shirt was covered with blood, and he breathed heavily. A moment after the woman uttered a low, wild cry, and, dashing through the crowd—her long black hair streaming behind her—she cast herself down by the prostrate man and sobbed,

"O cara mio! O Deos! is he dead? is he dead?"

"Who did this? Who stabbed this man?" demanded several voices, fiercely.

"No matter," answered the wounded man, faintly. "It was my own fault; I deserved it;" and, turning his face toward the weeping woman, he said, smiling, "Don't cry; don't go on so."

There was an ineffable tenderness in his voice, and something indescribably sweet in the expression of his face.

"O Deos!" cried the woman, kissing him passionately. "O cara mio! Say you will not die! Tell me you will not die!" And, tearing her dress with frantic strength, she tried to stanch the blood, which was rapidly forming a crimson pool around him.

The crowd meantime pressed so close that the man suffered for want of air, and begged to be removed. Several persons seized hold of him, and, lifting him from the floor, carried him out. The dark señorita followed close up, still pressing the fragments of her bloodstained dress to his wound.

Order was restored, and the music and dancing went on as if nothing had happened.

I had no desire to see any more of the evening amusements.

Next day I learned that the unfortunate man was dead. He was a stranger at San Luis, and refused to reveal his name, or make any disclosures concerning the affray. His last words were addressed to the woman, who clung to him with a devotion bordering on insanity. When she saw that he was doomed to die, the tears ceased to flow from her eyes, and she sat by his bedside with a wild, affrighted look, clutching his hands in hers, and ever and anon bathing her lips in the life-blood that oozed from his mouth.

"*I loved you—still love you better than my life!*"

These were his last words. A gurgle, a quivering motion of the stalwart frame, and he was dead!

At an examination before the alcalde, it was proved that the stabbing must have occurred before the affray became general. It was also shown that the young Mexican was unarmed, and had no acquaintance with the murdered man.

Who could have done it?

Was it the devil-woman? Was this a case of jealousy, and was the tall Texan the father of the murdered child?

Upon these points I could get no information. The whole affair, with all its antecedent circumstances, was wrapped in an impenetrable mystery. When the body was carried to the grave by a few strangers, including myself, the chief mourner was the half-breed woman—now a ghastly wreck. The last I saw of her, as we turned sadly away, she was sitting upon the sod at the head of the grave, motionless as a statue.

Next morning a vaquero, passing in that direction, noticed a shapeless mass lying upon the newly-spaded earth. It proved to be the body of the unfortunate woman, horribly mutilated by the wolves. The clothes were torn from it, and the limbs presented a ghastly spectacle of fleshless bones. Whether she died by her own hand, or was killed by the wolves during the night, none could tell. She was buried by the side of her lover.

Soon after these events, having completed my business in San Luis, I took passage in a small schooner for San Francisco, where I had the satisfaction in a few days of turning over ten thousand dollars to the Collector of Customs.

I never afterward could obtain any information respecting the two men mentioned in the early part of my narrative—the Colonel and Jack. No steps were taken by the authorities to arrest them. It is the usual fate of such men in California sooner or later to fall into the hands of an avenging mob. Doubtless they met with a merited retribution.

Eleven years have passed since these events took place. Many changes have occurred in California. The gangs of desperadoes that infested the state have been broken up; some of the members have met their fate at the hands of justice—more have fallen victims to their own excesses. I have meanwhile traveled in many lands, and have had my full share of adventures. But still, every incident in the "Dangerous Journey" which I have attempted to describe is as fresh in my mind as if it happened but yesterday.

OBSERVATIONS IN OFFICE.

I.

MY OFFICIAL EXPERIENCES.

THERE is something very fascinating in public office. The dignity of the position touches our noblest sympathies, and makes heroes and patriots of the most commonplace men. It is wonderful, too, how unselfish people become under the influence of this most potent charm. Every four years it becomes an epidemic. The passional attraction of office is felt throughout the length and breadth of the land. Many thousands of our best citizens visit the seat of government at the inauguration of a new president. A large proportion of them have faithfully served their country by contributing their time, talents, energies, and pecuniary resources to the success of the dominant party. But they don't want any thing; they have a natural repugnance to office; they merely come to look on, and pay their respects to the chief magistrate. If he deems it necessary to solicit their services for the common good, it is not for them, as patriotic citizens, to refuse. The seductive influences of official position may tend, perhaps, to quicken their perception of the grades of service in which their time could be most profitably spent; but modesty, after all, is their predominant trait. Indeed, for that matter, the general characteristic of great men is modesty, and where will you see so many notoriously great men as in Washington upon the advent of a new administration. The difficulty is to find a man who is not great. You may find many who are poor, some thriftless, and a few worthless, but none deficient in greatness. It must not be understood, however, that mercenary considerations have any connection with the charm which allures them thither. These excellent people—as in my own case, for example—are governed by motives of the purest and most exalted patriotism. Who is there so destitute of national pride—so indifferent to the welfare of his fellow-beings, that he does not desire to serve his country when he sees that she stands in need of his services?

The consideration of a per-diem allowance could not be wholly discarded, but I assure you, upon the veracity of a public officer, it had not the slightest influence upon me when I accepted the responsible position of Inspector General of Public Depositories. The Secretary of the Treasury—a gentleman in whom I had great confidence—required my services. I was unwilling, of course, to stand in the way of an efficient administration of the affairs of his department. The fact is, I had great personal respect for him, and was anxious to afford him all the assistance in my power. I do not pretend to say that the appointment of inspector general was destitute of attractions in itself, but they were not of a pecuniary character. The title had a sonorous and authoritative ring about it altogether different from the groveling jingle of filthy lucre—something that vibrated upon the higher chords of the soul.

An honorable ambition to serve one's country is one of the highest and most ennobling passions that can govern the human mind. To this may be attributed some of the greatest achievements which have given lustre to ancient and modern history. It has developed the greatest intellects of the Old and New World, and furnished the rising generation with illustrious models of unselfish devotion to the common welfare of mankind.

No wonder, then, that office possesses such extraordinary attractions. It is the cheapest way of becoming great. A man never before heard of outside of his village home—never before known to do any thing remarkable by his most intimate friends—never before suspected of possessing the least capacity for mental or manual labor of any kind whatsoever, may become, in the course of four-and-twenty hours, a topic of newspaper comment throughout the whole country—praised for virtues he never possessed, abused for vices to which he never aspired. An appointment places him prominently before the public. It shows the world that there was always something in him—whether whisky or sense matters little, since he has received the endorsement of the "powers that be."

To make a short story of it, I was obliged to accept the position. The party in power stood in need of my services. I could not refuse without great detriment to the country. This was many years since; and I beg to say that there is nothing in my journal of experiences bearing upon the present state of affairs. At great pecuniary sacrifice (that is to say, in a prospective sense, for I hadn't a dime in the world), I announced myself as ready to proceed to duty. In his letter of instructions, the Secretary of the Treasury was pleased to direct me to proceed to the Pacific coast, and carefully examine into the condition of the revenue service in that remote region. I was to see that the accounts of the collectors were properly kept and rendered; that the revenue laws were faithfully administered; that the valuation of imports was uniform throughout the various districts; whether any reduction could be made in the number of inspectors and aids to the revenue stationed within their limits, with a view to a more economical administration of the laws; whether the public moneys were kept in the manner prescribed by the Independent Treasury Act of August 6th, 1846; and what additional measures, if any, were necessary for the prevention of smuggling and other frauds upon the revenue, all of which I was to report, with such views as might be suggested in the course of the investigation for the promotion of the public interests. These were but a few of the important subjects of official inquiry upon which I was to enlighten the Department. I frankly confess that, when I read the instructions, and pondered over their massive proportions and severe tone of gravity, I was appalled at the immensity of the interests committed to my charge. A somewhat versatile career, during which I had served before the mast in a whaler, studied medicine, hunted squirrels in the backwoods, followed the occupation of ferry-keeper, flat-boat hand, and short-hand writer, had not fitted me particularly for this sort of business. What did I know about the forms of accounts current, drawbacks, permits, entries, appraisements, licenses, enrollments, and abstracts of imports and exports? What reliable or definite information was I prepared to give to collectors of customs in reference to schedules and sliding scales? What hope was there that I could ever get to the bottom of a fraud upon the revenue service, when I had but a glimmering notion of the difference between fabrics of which the component parts were two thirds wool, and fabrics composed in whole or in part of sheet-iron, leather, or gutta-percha?

As for inspectors of customs, how in the world was an agent to find out how many inspectors were needed except by asking the collector of the district, who ought to know more about it than a stranger? But if the collector had half a dozen brothers, cousins, or friends in office as inspectors, would it not be expecting a little too much of human nature to suppose he would say there were too many in his district? I reflected over the idea of asking one of these gentlemen to inform me confidentially if he thought he could dispense with a dozen or so of his relatives and friends without detriment to the public service, but abandoned it as chimerical. Then, to go outside and question any disinterested member of the community on this subject seemed equally absurd. Who could be said to be disinterested when only a few offices were to be filled, and a great many people wished to fill them? I would be pretty sure to stumble upon some disappointed applicant for an inspectorship, or, worse still, upon a smuggler. It is a well-ascertained fact that disappointed applicants for office are always opposed to the fortunate applicants, and smugglers, as a general rule, have a natural antipathy to inspectors of customs.

There was another serious duty imposed upon me—to ascertain the character and standing of all the public employés, their general reputation for sobriety, industry, and honesty, and to report accordingly. Here was rather a delicate matter—one, in fact, that might be productive of innumerable personal difficulties. Having no unfriendly feeling toward any man, and attaching a fair valuation to life, I did not much relish the notion of placing any man's personal infirmities upon the official records. If a public officer drank too much whisky, it was certainly a very injurious practice, alike prejudicial to his health and morals; but where was to be the gauge between too much and only just enough? No man likes to have his predilection for stimulating beverages made a matter of public question, and the gradations between temperance and intemperance are so arbitrary in different communities that it would be a very difficult matter to report upon. I have seen men "sociable" in New Orleans who would be considered "elevated" in Boston, and men "a little shot" in Texas who would be regarded as "drunk" in Maine. It is all a matter of opinion. No man is ever drunk in his own estimation, and whether he is so in the estimation of others depends pretty much upon their standard of sobriety. With respect to honesty, that was an equally delicate matter. What might be considered honest among politicians might be very questionable in ordinary life. I once knew of a public officer who had been charged with embezzling certain public moneys. There was no doubt of the fact, but he fought a duel to prove his innocence. In one respect, at least, he was honest—he placed a fair valuation upon his life, which was worth no more to the community than it was to himself. I did not think an ordinary per-diem allowance would be sufficient to compensate for maintaining the public credit by such tests as this, especially as there were nearly two hundred public offices to be examined; but it seemed nothing more than reasonable that the laws should be administered by sober and honest men, and, upon the whole, I could not perceive how this unpleasant duty could be avoided.

The Department furnished me with a penknife, a pencil, several quires of paper, and a copy of Gordon's Digest of the Revenue Laws. This was my outfit. It was not equal to the outfit of a minister plenipotentiary, but there was a certain dignity in its very simplicity. To be the owner of a fine Congressional penknife, a genuine English lead-pencil, paper *ad libitum*, and Gordon's Digest, was no trifling advance in my practical resources. I looked into the Digest, read many of the laws, and became satisfied that the Creator had not gifted me with any capacity for understanding that species of writing. For Mr. Gordon, who had digested those laws, I felt a very profound admiration. His powers of digestion were certainly better than mine. I would much rather have undertaken to digest a keg of spike nails. The Act of March 2, 1799, upon which most of the others were based, was evidently drawn with great ability, and covered the whole subject. Like a Bœotian fog, however, it covered it up so deep that I don't think the author ever saw it again after he got through writing the law. Whenever there was a tangible point to be found, it was either abolished, or so obscured by some other law made in conformity with the progress of the times that it became no point at all; so that, after perusing pretty much the whole book, and referring to Mayo's Compendium of Circulars and Treasury Regulations, I am free to confess the effect was very decided. I knew a great deal less than before, for I was utterly unable to determine who was right—Congress, Gordon, Mayo, or myself.

Under these circumstances, it will hardly be a matter

of surprise that serious doubt as to my capacity for this service entered my mind. Perhaps, in the whole history of government offices, it was the first time such a doubt ever entered any man's head upon receiving an appointment, and I claim some credit for originality on that account. The position was highly responsible; the duties were of a very grave and important character, bordering on the metaphysical. Now, had I been requested to visit Juan Fernandez, and report upon the condition of Robinson Crusoe's castle, or ascertain the spot in which he found the footprint in the sand, or describe for the benefit of science the breed of wild goats descended from the original stock—had these questions been involved in my instructions, or had I been appointed to succeed Sancho Panza in the government of Nantucket (which I verily believe was the island referred to by Cervantes), I could have had no misgivings of success. But this awful thing of abstracts and accounts current; this subtile mystery of appraisements, appeals, drawbacks, bonds, and bonded warehouses; this terrible demon of manifests, invoices, registers, enrollments, and licenses; this hateful abomination of circulars on refinéd sugar, and fabrics composed in whole or in part of wool; this miserable subterfuge of triplicate vouchers and abstracts of disbursements, combined to cast a gloom over my mind almost akin to despair. The question arose, would it not be the most honorable course to return the commission to the Secretary of the Treasury, and confess to him confidentially, as a friend, that I thought he would render the country greater service by appointing a more suitable agent? But then there was the per-diem allowance, a very snug little sum, much needed at the time; and there was the honor of the position—a pillar in the federal structure; and then the advantage of travel, and the charm of becoming at once famous in the national records. Besides, it might be considered disrespectful to say to the Secretary of the Treasury—a gentleman from an interior state, who had no experience in commerce or public finances—that I had no experience in these things myself, and doubted my capacity to do justice to the government. Might he not regard such a confession in the light of a personal reflection?

After all, I thought it would be as well perhaps to try my hand at the business. Many a man never finds out that he is great in some particular line till he tries his hand. I have at this moment in my eye at least half a dozen senators of the United States who I verily believe would make excellent butchers, bakers, blacksmiths, and carpenters, if they only knew it. I am acquainted with some that would be an ornament to any court of justice as public criers, and not a few who would make capital hands at playing quoits and pitch-penny. In short, I know many men occupying these positions who would succeed even better in other branches of industry than in the capacity of statesmen; but the misfortune is, they are not aware of the fact, and never can be persuaded to believe it. Very few men understand what they are good for till some adventitious circumstance occurs to develop their latent and peculiar talents. In this view, it might be that I was a capital hand at revenue business, though, to tell the truth, I had never collected any revenue worth mentioning on my own account; and what experience I had had in depositories was confined to my own pockets, which seldom retained the sums deposited in them over twelve hours, if so long as that.

A transcript from my official reports will convey some idea of my labors under the complicated instructions issued to me at various intervals from Washington.

The first has reference to the general subject of smuggling, and proposes the removal of the Custom-house from San Francisco to Bear Harbor, near Cape Mendocino. This was addressed to his Honor the Secretary of the Treasury:

"Sir,—If Bear Harbor is eligible for any purpose in the world, it is for a port of entry and a custom-house.

Not that there are any inhabitants there at present, or in the vicinity, except Indians, bears, elk, deer, and wild-cats; not that any vessels ever come in there, or ever will, perhaps, but as a guard against smuggling. You know, sir, from the experience of collectors from Passamaquoddy Bay to Point Isabel, and from San Diego to the Straits of Fuca, that smuggling must be going on somewhere, else why is the Treasury Department flooded with applications for an increase of inspectors? Even senators and members of Congress unite in the opinion that a great deal of smuggling is perpetrated on remote and isolated parts of our coast, for they are always recommending some friend in whom they have confidence to keep a guard upon the revenue at such places. One would think that smugglers would rather pay duties and take their wares into a good market, than put them ashore where there are no inhabitants, and transport them at double the risk and cost to some place where they are wanted. If they must enjoy the pleasure of violating the law at all, would it not pay better to smuggle their wares directly into the principal cities, as New York or San Francisco, for example? I know that in the former place they incur some risk of detection from night inspectors, who are supposed to be always on the look-out about the wharves after dark; but in San Francisco the night inspectors have been abolished on account of the soporific effects of the climate. Several of them fell asleep directly after receiving their appointments, and never woke up, except on pay-day, during the entire term of their service.

"For some years, at least, one collector of customs could perform all the duties that might be required of him at Bear Harbor. No doubt the dullest and laziest politician in the entire state could be hired to occupy the position at three thousand dollars per annum. The collector at the city of Gardner, which consists of two small frame shanties and a pig-pen, situated at the mouth of the Umpqua River—where shipwreck is almost absolutely certain in case a vessel attempts to enter—receives only a thousand dollars per annum. Sir, it can not be expected that a gentleman more than ordinarily gifted with valuable traits of character can be obtained for so small a sum. Government is compelled to pay for the services of active and intelligent collectors at the ports of Benicia, Sacramento, Stockton, Monterey, San Pedro, and San Diego, three thousand dollars a year each. If they were at all conspicuous for idleness, it is impossible to conjecture what it would cost to obtain their services; but the amount of labor performed by these gentlemen (who, by the way, are all very excellent persons, and for whom I entertain great personal respect) is almost incredible. At Benicia the duties of the office are absolutely onerous. From one to two vessels a year enter that port with coals from Cardiff, which are deposited at the dépôt of the Pacific Steam-ship Company. Upon these coals the duties have to be computed and accounts rendered to the Department, besides which he is compelled to keep an accurate account of his own salary. For all this he is only allowed the occasional services of one inspector, whereas he ought to be allowed three. If they were gentlemen of a lively temperament, they would at least give something of vitality to the present deserted appearance of the port. I have known a smaller number than that to produce a considerable sensation in the public streets of other cities. A great deal of trouble to the Benicia collector might be saved if the two Cardiff vessels per annum were permitted to enter at San Francisco on their way up.

"At Sacramento the duties of the collector are still more arduous. Indeed, it is a matter of surprise that any man can be found to undertake them at three thousand dollars a year. A vessel with foreign goods entered this port in 1849, since which period some six or eight consecutive collectors have been anxiously awaiting the arrival of another. The most remarkable part of it is, that the other vessel has never yet arrived. Upon a review of the facts, I think that any person of a less sanguine temperament than a collector of customs would have long since given up the hope of obtaining any public revenue from this source. Somehow all the vessels have a habit of stopping at San Francisco, paying duties there, discharging their cargoes for interior transportation, and going about their business, which must be a constant subject of mortification to the Sacramento collectors. I have known respectable gentlemen who occupied this position to be denied over twenty-five dollars a month for office-rent, after it had ranged for years at two or three hundred—even denied the services of a deputy or clerk, and actually compelled to make out their own pay accounts!

"And yet these officers are required to attend at primary meetings, conventions, and legislative assemblages, and keep the party all right, when there may be a complication of difficulties between the various aspirants for the Senate of the United States, utterly impossible to settle except by electing them all.

"At Stockton the case is still harder. I never knew a collector there to have any thing at all to do, except to keep the run of his office-rent and salary, which, in justice it must be said, is a branch of public duty always faithfully performed. Yet this officer is expected to pass the time agreeably year after year on a miserable pittance of three thousand dollars, without even the hope of ever seeing a dutiable cargo landed upon the wharves of the city. I do not believe that the most sanguine gentleman that ever held that position aspired to any thing of greater commercial value than a flock of sheep supposed to be on the way from Mexico, and for the capture and confiscation of which two inspectors were for many years stationed at the Tejon Pass, about three hundred miles from Stockton. But even the hope of seizing these sheep or their descendants has been blasted since Congress abolished the duties on stock; and now the collector, to protect the revenue, must fail unless he succeeds in getting hold of a box of contraband articles that it is supposed certain parties in San Francisco are awaiting an opportunity to send up, either by the steam navigation line or some of the small sailing craft that ply on this route. As this box of goods has been expected ever since 1852, the prospect of its appearance and seizure is becoming more favorable every year. If there was a surveyor stationed at the mouth of the San Joaquin—say in the city called 'the New York of the Pacific'—the chances of seizure would be greatly augmented. There is a surveyor of customs at Nisquelly, in the Territory of Washington, and another at Santa Barbara, who might render some aid by the transmission of secret information. I do not know what has become of the surveyor at Pacific City, near the mouth of the Columbia. The last time I saw him he was engaged in the performance of his official functions in the tin business at Oregon City, the City of Pacific having been discontinued about two years previously in consequence of a lack of inhabitants.

"At Monterey the amount of hardship endured by the collector is absolutely incredible. Not only is he furnished with an indifferent government house to live in, which costs an annual outlay of several hundred dollars to keep it from falling to pieces, and thereby crushing himself and assistants beneath the ruins, but he is required to look after two inspectors, who are appointed to aid him in protecting the coast from the nefarious operations of smugglers. Besides this, it is supposed that a mysterious vessel has been hovering around the Bay of Monterey ever since 1852, with an assorted cargo of bar fixtures, billiard balls, whisky, nine-pins, cards, cotton handkerchiefs, boots, bowie-knives, and revolvers, upon a considerable portion of which duties have never been paid. This vessel is no doubt awaiting an opportunity to land these articles in violation of law, and to the great detriment of public morals and serious loss to the treasury. The collector is expected to be present or within reach of a telegraphic dispatch whenever she makes her appearance;

and it is further expected that he will not flinch from his duty even should she prove to be the Flying Dutchman or the Wizard of the Seas.

"At San Pedro the coasting steamer Senator touches for grapes and passengers some half a dozen times a month, and the collector is expected to keep a record of that vessel's arrivals and departures; also the range of Captain Banning's paddle-wheeled steam skiff Medora, six scows, and several fishing smacks. In addition to these onerous duties, it devolves upon him to keep his own pay account, and see that the light does not stop burning of nights in the public light-house on Point Conception, without any money to pay the keeper and assistant except such casual remittances as may be made once or twice in the course of as many years. I knew one light-house keeper who stood by the light manfully for a whole year, and finally had to sell his chance of pay for the means of subsistence. Some of the light-house keepers, indeed, are supposed to live on whale-oil, the Board in Washington being evidently under the impression that oil is a light article of diet, upon which men will not be apt to go to sleep. Another reason, perhaps, for the remissness with which their salaries generally arrive is that their stations are generally not densely populated with voters, or, in fact, with any thing but sheep and rabbits. I have a person in my eye whom I would like to recommend for the collectorship at San Pedro whenever the present incumbent may think proper to resign. By the way, the latter is a very clever and estimable gentleman, to whom I intend not the slightest disrespect in thus referring to his office; but there are peculiar qualifications for every position in life, and the individual to whom I refer possesses some very remarkable advantages over the generality of custom-house officers; that is to say, he can sleep on his desk in the midst of the direst confusion; is never known to be in a hurry; thinks no more of time than he does of eternity, or any thing else; and invariably postpones till to-morrow what most people would deem of vital importance to be done to-day. His work is generally in arrears, but will be all right—*poco tiempo!*

"At San Diego the same burdensome and oppressive state of things exists. The Custom-house is an old military building, with a roof that falls to pieces every winter, and a set of doors and windows through which both wind and rain have free access. The only article of public property about the premises that yet sticks together is a tremendous iron safe, in which the revenue is going to be kept—as soon as it is collected. Even this is getting rusty for want of use. The books have an ancient and fish-like aspect; and a public shovel, that is used to clear the mud away from the door whenever a vessel is seen in the offing, is going away year after year, and will eventually be reduced to a broken handle. This office is accessible by means of a boat, though in bad weather the deputy prefers to reside in an old hulk that lies at anchor in the bay. The building is eligibly located in a chapparal of prickly pears, within about five miles of Old Town, or, properly speaking, the beautiful city of San Diego. Mexican stock were formerly imported into this district, but, having been made free by act of Congress, the collector is left destitute of occupation, and is compelled to seek business and society in various parts of the state. Now and then, however, he is supposed to take a look at his pay account, and see that the public light on the Point keeps burning of nights, notwithstanding the roof has been blown off. As government refuses to furnish him with rain-water to drink, he is compelled, whenever his official duties call him to the port of entry, to hitch up his buggy and travel five miles to the city of San Diego every time he is thirsty. Indeed, so parsimonious is the Department becoming of late, that it will not even allow him a deputy or clerk at public expense, although there has been one there for years. I look upon this as a very severe course of discipline to impose upon any gentleman whose services are presumed to be worth three thousand dollars per annum, and would recommend that he should at least be allowed a bottle of whisky.

"All these are examples of the manner in which executive patronage may be enlarged without inconvenience to commerce or obstruction to navigation. If it were not for the collectorships, what would the delegation in Congress have to make up the complement of their indebtedness to partisan politicians? and if one delegation were denied this privilege, how could accounts be settled with fellow-members similarly situated in other states? An inspector of customs, at a compensation of five hundred dollars a year (for there is nothing to do), would of course answer the requirements of commerce at any of these ports; but then what sort of an office would that be to offer to the owner of one or more members of the Legislature? It would be especially severe at Bear Harbor, where there will be no coffee-houses, billiard saloons, or other places of amusement for some time.

"In view of these suggestions being urged upon Congress by the heads of the departments, I would mention that, in the temporary absence of government buildings at Bear Harbor, a number of chapadens, or brush tents, at present occupied by Indians, can be leased for a term of years at a rate of rent not exceeding from five hundred to a thousand dollars each per month. The very best of them can be had for less than the rent paid for the Union Street Bonded Warehouse in San Francisco, toward the building of which government loaned seventy-two thousand dollars as an advance of rent, and paid, by way of interest on the capital, for four years, two hundred and eighty-eight thousand dollars; after which, upon the united representation of twenty influential merchants, a collector and deputy collector of customs, and a special agent, that the premises were only worth about fourteen thousand per annum, it paid one hundred and ten thousand more to abrogate the contract, and as a solemn warning to all private individuals and public officers not to attempt such a speculation as that again. The chapaden of the chief digger, To-no-wauka, could be purchased in fee simple for less than twenty-eight thousand dollars, which was the exact amount annually expended for the rent of the United States Court-rooms at San Francisco until my friend Yorick, the government agent, reduced it to ten thousand, after which, of course, he was removed.

"As an additional protection to the revenue, I would suggest that a revenue cutter be stationed at Bear Harbor, modeled after the fashion of a large wash-tub, which would be but a slight improvement upon the sailing capacity of the three cutters now stationed on the Pacific coast. The masts might be constructed out of large tin dippers inverted, in the bowls of which marines could be stationed to keep a look-out for smugglers. Spare blankets would answer for the sails, and a large carving-knife run out at the stern would serve admirably to steer by. In order that there might be no danger of missing the way during dark nights from any variation in the compass, it would be well, perhaps, to abandon the compass altogether, and send a boat ahead with a light, to point out where the rocks and smugglers might be found. There being no vessels to catch at Bear Harbor, no inconvenience would result from the fact that such a cutter would be as well calculated to lie at anchor as the cutter Marcy at San Francisco, which has been known to pursue several vessels for infractions of the revenue laws, but never to catch any of them. I attribute this not to any want of zeal on the part of the officers, but partly to the superior speed of the runaway vessels, and partly to the fact that the Marcy is obliged to lie at anchor for six months in the year in the Bay of San Francisco for want of other occupation. The remaining six months she necessarily spends in the Straits of Carquinas, near Benicia, in order to get rid of the barnacles that accumulate on her bottom during the term of her sedentary career below.

"If exception should be taken to this precedent on

the ground that a revenue cutter may sometimes really be wanted at a port of entry where there is some commerce, surely none will be taken to the cutter Lane, stationed within the mouth of the Columbia River. For the officers of this cutter I entertain the most sincere respect; but if she has ever been known to chase any thing larger than wild ducks, the fact must have been hushed up from motives of public policy. It has certainly not been a matter of general comment. About one vessel with dutiable merchandise enters the Columbia in the course of half a dozen years, and certainly all sailing vessels have difficulty enough in getting in, without attempting to run away after they come to an anchor. Indeed, I don't know where they would run to unless it might be over the Cascades, and through the Dalles to Walla Walla, or up to Oregon City on the Willamette River, where the flour-mills of Abernethy & Co. would soon grind them to pieces. To suppose that they would undertake to run away before they get over the bar is to suppose that they might just as well stay away altogether, and thereby avoid the risk of shipwreck in addition to the remote possibility of being captured by a revenue cutter. The officers condemned to this station have my most ardent sympathies. It generally rains at Astoria between two and three hundred days every year, the consequence of which is, that the whole country and every thing in it has a mildewed appearance. Already I can fancy that barnacles are growing on the beards of these gentlemen; that their skin is becoming slippery and green; their eyes sharkish in expression, from a constant habit of looking out for smugglers that never can be within five hundred miles; that the habit of pulling ashore in the boats and back again; 'making it so' when four and eight bells are announced; looking up at the mast-head and then down again; going below and reading the same old newspaper, and coming up again; turning in and taking a nap, and turning out when the nap is ended; exercising their quadrants by an occasional peep at the heavenly bodies; eating three scanty and melancholy meals a day; doing all this and never doing any thing else, unless it may be to superintend the patching of an old sail which has rotted to pieces, or the splicing of an old rope to keep the blocks from falling down on their heads, will eventually so wear upon their mental and physical resources as to drive them all mad. Should it ever be the misfortune of any suspicious character to fall into the hands of these gentlemen, I have no doubt he will have reason to regret it during the brief period of his existence; for they will certainly cut him to pieces with their swords, or blow him to fragments out of one of the public guns, on the general principle that, being paid for doing something, they ought to do it as soon as possible.

"The revenue cutter at Puget's Sound, familiarly known as the 'Jeff Davis,' finds occasional occupation in chasing porpoises and wild Indians. It is to be regretted that but little revenue has yet been derived from either of these sources; but should she persist in her efforts, there is hope that at no distant day she may overhaul a canoe containing a keg of British brandy—that is to say, in case the paddles are lost, and the Indians have no means of propelling it out of the way.

"These vessels, in addition to their original cost, which was not cheap considering their quality and sailing capacity, require an expenditure of some forty or fifty thousand dollars a year for repairs, rigging, pay of officers and men, subsistence, etc., as also for powder to enable the officers to kill ducks and salute distinguished people that visit these remote regions. Now and then they run on the rocks in trying to find their way from one anchorage to another, in which event they require extra repairs. As this is for the benefit of navigation, it should not be included in the account. They generally avoid running on the same rock, and endeavor to find out a new one not laid down upon the charts—unless, perhaps, by some reckless fly—in order that other vessels may enjoy the advantage of additional experience. The beauty of Bear Harbor in this respect is, that a revenue cutter could run on a new rock every day in the year, so that, by designating its exact location on the chart, there would be three hundred and sixty-five rocks per annum to be avoided by vessels entering the harbor.

"Some military protection would probably be required there for several years to come, in order to protect the citizens from the attacks of grizzly bears. I would suggest that a post be established on some eligible point, and comfortable quarters erected for the officers and soldiers. While these quarters are in progress of erection, it might be well to station a large rooster in the top of a neighboring tree to give warning of the approach of the enemy. As Rome was saved in one way, so might Bear Harbor be saved in another. Should it become necessary to abandon them, the citizens will no doubt be willing to purchase them at public auction.

"I do not know what the military quarters at Fort Miller are going to do, but the last time I saw them they looked very sorry they had ever been built. The same may be said of the quarters at Benicia, Fort Tejon, and San Diego, which goes to prove the transitory character of military operations. So long as our army goes about the country dropping down beautiful little cities, we in the line of civil life can certainly have no objection. As expense is no object, perhaps, to the War Department, I would suggest that there is a very rugged point of rocks near the entrance of Bear Harbor, upon which a friend of mine has located a claim that he is willing to sell for military purposes for the sum of one hundred and fifty thousand dollars. It commands a fine view of the ocean, and abounds in mussels and albicores; besides which, it is cheaper and uglier than Lime Point at the entrance of the Golden Gate, and would not require near so much writing to make the purchase satisfactory to the public.

"For a few years, during the infancy of the community, it may be necessary for some enterprising citizen to borrow from government one hundred thousand dollars at six per cent. per annum, in consequence of the high rates of interest in California. There will be no difficulty in doing this, I apprehend, if he have influence at court. A precedent may be found in the case of the Folsom estate, against which judgment had been obtained, and an execution placed in the hands of the marshal. Private parties found it to their advantage to step in, purchase a portion of the property, pay a portion of the debt, and, upon giving satisfactory security, assume the remainder, amounting to a hundred thousand dollars, at six per cent. It may be a little irregular to favor particular parties in this way, but then public money had better be bringing six per cent. than lying idle in the treasury; and besides, when it is found necessary to issue treasury notes in order to carry on the government, they bring a premium, and there is a gain to that extent over the ready cash. If all the public money was loaned out at six per cent., and all the private money that might be necessary borrowed at five, of course the financial condition of the treasury would be one per cent. better per annum.

"After these things were done, and the business of Bear Harbor placed upon a permanent footing, private instructions might be issued to the collector of customs to go out and stump the state in behalf of the great principles of national economy. Experience would enable him to stand firmly upon the broad platform of public integrity; and when he addressed the multitude, he could dwell feelingly on the sublime doctrine of earlier days—'Millions for defense, but not a cent for tribute!' He could put his hand upon his brow, and solemnly declare that, so long as he was gifted with the light of intellect to comprehend the sound doctrines of public policy bequeathed to us by our forefathers, he would stand by the laws and the Constitution. He could put his hand upon his heart, and call upon the people to witness that he, for one, had ever remained true to first principles. He could put his hand upon his stomach, and avow, from the bottom of his soul, that he conscientiously indorsed the measures of the prevailing party. He could put his hand upon his pocket, and affirm in all sincerity that he went heart and hand with the reigning powers on all the great questions of the day. And, having fully delivered himself on these various points, he could wind up with an anecdote from the Schildburghers. When the wise men of Schilda undertook to build their grand council-house, they carried down on their backs from the top of a high hill a large number of heavy logs. In moving the last log, it fell out of their hands and rolled to the bottom of the hill. 'Don't you see,' said the town fool, 'if you had started them all in the same way, they would have rolled down of their own accord?' which they admitted was true, and accordingly carried all the logs up to the top of the hill again, and then rolled them down. So, if the people don't like this party, they can roll in another just as good. Your obedient servant, etc."

In my next chapter of experiences I propose giving a succinct account of the great Port Townsend Controversy. This cost me more trouble than all my other experiences together, and came very near costing me my life.

II.

THE GREAT PORT TOWNSEND CONTROVERSY, SHOWING HOW WHISKY BUILT A CITY.

Few persons who have visited the Pacific coast of late years are ignorant of the fact that the city of Port Townsend is eligibly situated on Puget's Sound, near the Straits of Fuca; and none who have seen that remarkable city can hesitate a moment to admit that it is a commercial metropolis without parallel.

Port Townsend is indeed a remarkable place. I am not acquainted with quite such another place in the whole world. It certainly possesses natural and artificial advantages over most of the cities known in the Atlantic States or Europe. In front there is an extensive water privilege, embracing the various ramifications of Puget's Sound. Admiralty Inlet forms an outlet for the exports of the country, and Hood's Canal is an excellent place for hoodwinking the revenue officers. On the rear, extending to Dunganess Point, is a jungle of pine and matted brush, through which neither man nor beast can penetrate without considerable effort. This will always be a secure place of retreat in case of an invasion from a war-canoe manned by Northern Indians. With regard to the town itself, it is singularly picturesque and diversified. The prevailing style of architecture is a mixed order of the Gothic, Doric, Ionic, and Corinthian. The houses, of which there must be at least twenty in the city and suburbs, are built chiefly of pine boards, thatched with shingles, canvas, and wooden slabs. The palace and out-buildings of the Duke of York are built of drift-wood from the saw-mills of Port Ludlow, and are eligibly located near the wharf, so as to be convenient to the clams and oysters, and afford his maids of honor an opportunity of indulging in frequent ablutions. There is somewhat of an ancient and fish-like odor about the premises of his highness, and it must be admitted that his chimneys smoke horribly, but still the artistic effect is very fine at a distance. The streets of Port Townsend are paved with sand, and the public squares are curiously ornamented with dead horses and the bones of many dead cows, upon the beef of which the inhabitants have partially subsisted since the foundation of the city. This, of course, gives a very original appearance to the public pleasure-grounds, and enables strangers to know when they arrive in the city, by reason of the peculiar odor, so that, even admitting the absence of lamps, no person can fail to recognize Port Townsend in the darkest night. When it was a port of entry under the laws of the United States, there was a collector of customs stationed in a small shanty on the principal wharf, whose business it was to look out for smugglers, and pay the salary of an inspector, who owns some sheep on San Juan Island, and holds joint possession of that disputed territory with the British government. The collector of customs, being unable to attend to the many important duties that devolved upon him without assistance, was allowed two boatmen, whose duty it was to put him on board of suspicious vessels in the offing, and one of whom, by virtue of a special commission, was ex-officio deputy collector, and made up the accounts of the district.

The principal luxuries afforded by the market of this delightful sea-port are clams, and the carcasses of dead whales that drift ashore, by reason of eating which the inhabitants have clammy skins, and are given to much spouting at public meetings. The prevailing languages spoken are the Clallam, Chenook, and Skookum-Chuck, or Strong Water, with a mixture of broken English; and all the public notices are written on shingles with burnt sticks, and nailed up over the door of the town-hall. A newspaper, issued here once every six months, is printed by means of wooden types whittled out of pine knots by the Indians, and rubbed against the bottom of the editor's potato pot. The cast-off shirts of the inhabitants answer for paper. For the preservation of public morals, a jail has been constructed out of logs that drifted ashore in times past, in which noted criminals are put for safe keeping. The first and last prisoners ever incarcerated in that institution were eleven Northern Indians, who were suspected of the murder of Colonel Ehy at Whidbey's Island. As the logs are laid upon sand to make the foundation secure, the Indians, while rooting for clams one night, happened to come up at the outside of the jail, and finding the watchman, who had been placed there by the citizens, fast asleep, with an empty whisky bottle in the distance, they stole his blanket, hat, boots, and pipe, and bade an affectionate farewell to Port Townsend.

The municipal affairs of the city are managed by a mayor and six councilmen, who are elected to office in a very peculiar manner. On the day of election, notice having been previously given on the town shingles, all the candidates for corporate honors go up on the top of the hill back of the water-front, and play at pitch-penny and quoits till a certain number are declared eligible; after which all the eligible candidates are required to climb a greased pole in the centre of the main public square. The two best then become eligible for the mayoralty, and the twelve next best for the common council. These fourteen candidates then get on the roof of the town-hall and begin to yell like Indians. Whoever can yell the loudest is declared mayor, and the six next loudest become the members of the common council for the ensuing year.

While I had the misfortune to be in public employ (and for no disreputable act that I can now remember), it became my duty to inquire into the condition of the Indians on Puget's Sound. In the course of my tour I visited this unique city for the purpose of having a "wa-wa" with the Duke of York, chief of the Clallam tribe.

The principal articles of commerce, I soon discovered, were whisky, cotton handkerchiefs, tobacco, and cigars, and the principal shops were devoted to billiards and the sale of grog. I was introduced by the Indian Agent to the Duke, who inhabited that region, and still disputed the possession of the place with the white settlers. If the settlers paid him any thing for the land upon which they built their shanties it must have been in whisky, for the Duke was lying drunk in his wigwam at the time of my visit. For the sake of morals, I regret to say that he had two wives, ambitiously named "Queen Victoria" and "Jenny Lind;" and for the good repute of Indian ladies of rank, it grieves me to add that the Queen and Jenny were also very tipsy, if not quite drunk, when I called to pay my respects.

The Duke was lying on a rough wooden bedstead, with a bullock's hide stretched over it, enjoying his ease with the ladies of his household. When the agent informed him that a Hyas Tyee, or Big Chief, had called to see him with a message from the Great Chief of all the Indians, the Duke grunted significantly, as much as to say "that's all right." The Queen, who sat near him in the bed, gave him a few whacks to rouse him up, and by the aid of Jenny Lind succeeded, after a while, in getting him in an upright position. His costume consisted of a red shirt and nothing else, but neither of the royal ladies seemed at all put out by the scantiness of his wardrobe. There was something very amiable and jolly in the face of the old Duke, even stupefied as he was by whisky. He shook me by the hand in a friendly manner, and, patting his stomach, remarked, "Duke York belly good man!"

Of course I complimented him upon his general reputation as a good man, and proceeded to make the usual speech, derived from the official formula, about the Great

THE DUKE OF YORK, QUEEN VICTORIA, AND JENNY LIND.

Chief in Washington, whose children were as numerous as the leaves on the trees and the grass on the plains.

"Oh, dam!" said the Duke, impatiently; "him send any whisky?"

No; on the contrary, the Great Chief had heard with profound regret that the Indians of Puget's Sound were addicted to the evil practice of drinking whisky, and it made his heart bleed to learn that it was killing them off rapidly, and was the principal cause of all their misery. It was very cruel and very wicked for white men to sell whisky to the Indians, and it was his earnest wish that the law against this illicit traffic might be enforced and the offenders punished.

"Oh, dam!" said the Duke, turning over on his bed, and contemptuously waving his hand in termination of the interview—"dis Tyee no 'count!"

While this wa-wa, or grand talk, was going on, the Queen put her arms affectionately around the Duke's neck, and giggled with admiration at his eloquence. Jenny sat a little at one side, and seemed to be under the combined influence of whisky, jealousy, and a black eye. I was subsequently informed that the Duke was in the habit of beating both the Queen and Jenny for their repeated quarrels, and when unusually drunk was not particular about either the force or direction of his blows. This accounted for Jenny's black eye and bruised features, and for the alleged absence of two of the Queen's front teeth, which it was said were knocked out in a recent brawl.

Some months after my visit to Port Townsend, in writing a report on the Indians of Puget's Sound, I took occasion to refer to the salient points of the above interview with the Duke of York, and to make a few remarks touching the degraded condition of himself and tribe, attributing it to the illegal practice on the part of the citizens of selling whisky to the Indians. I stated that his wigwam was situated between two whisky-shops, and that the Clallams would soon be reduced to the level of bad white men in Port Townsend, "which, to say the least of it, was a very benighted place." The report was printed by order of Congress, though I was not aware of that fact till one day, sitting in my office in San Francisco, I received a copy of the "Olympia Democrat" (if I remember correctly), containing a series of grave charges against me, signed by the principal citizens of Port Townsend. I have lost the original documents, but shall endeavor to supply the deficiency as well as my memory serves. The letter was addressed to the "United States Special Agent," and was substantially as follows:

"Sir,—The undersigned have read your official report relative to the Indians of Puget's Sound, and regret that you have deemed it necessary to step so far aside from the line of your duty as to traduce our fair name and reputation as citizens of Port Townsend. You will pardon us for expressing the opinion that you might have spent your time with more credit to yourself and benefit to the government.

"Sir, it may be that on the occasion of your visit here the Duke of York and his wives were drunk; but the undersigned are satisfied, upon a personal examination, that neither Queen Victoria nor Jenny Lind suffered the loss of two front teeth, as you state in your report; and they are not aware that Jenny Lind's eyes were ever blacked by the Duke of York, nor do they believe it, although you have thought proper to make that statement in your report.

"The undersigned do not pretend to say that there is no whisky sold in Port Townsend; but they deny, sir, that you ever saw any of them drunk, or that the citizens of Port Townsend, as a class, are at all intemperate. On the contrary, they claim to be as orderly, industrious, and law-abiding as the citizens of any other town on the Pacific coast or elsewhere.

"Sir, it is scarcely possible that you can have forgotten so soon the marked kindness and hospitality with which you were treated by the citizens of this place during your sojourn here; and now the return you make is to blacken the reputation of our thriving little town, and endeavor to destroy our future prospects. You are, of course, at liberty to choose your own line of travel, but if ever you visit Port Townsend again, we can assure you, sir, you will enjoy a very different reception. Had you confined your misstatements to the Indians, we might have excused it on the ground that it is not customary for public officers to adhere strictly to facts in their reports; but when you go entirely out of your way, and commit such an unprovoked attack upon our character, we feel bound to set ourselves right before the world.

"In charity, we can only suppose that you have been grossly deceived in your sources of information; yet, when you profess to have witnessed personally the evil effects of whisky in Port Townsend, and go so far as to pronounce it 'a benighted place,' we can not evade the conclusion that you must have had some experience in what you say you witnessed; either that, or you deliberately committed a base slander upon the citizens of this place. Although the undersigned consider themselves included in your sweeping assertion, it can not have escaped your memory, sir, that on the occasion of your visit to Port Townsend you found them engaged in their peaceful avocations as useful and respectable members of society; and they positively deny that any of them have ever sold whisky to the Indians, or committed the crime of murder.

"Sir, the undersigned have made inquiry into that portion of your report in which you state that no less than six murders were committed here during the past year, and can only find that two were committed, and neither of them by citizens of this place. The conclusion, therefore, to which the undersigned are forced, is, that you were at a loss for something to say, and invented at least four murders for the purpose of contributing to the interest of your report.

"Sir, when a respectable community are engaged in trying to make an honest living, we think it hardly fair that you, as a government agent, should come among them, and, without cause or provocation, slander their character and injure their reputation. We therefore enter our solemn protest against the unfounded charges made in your report, and respectfully recommend that in future you confine yourself to your official duties.

"(Signed), J. Hodges, B. Punch, T. Thatcher, B. Fletcher, Warren Hastings, Wm. Pitt, J. Fox, E. Burke, and eleven others."

Here was a serious business. I can assure the reader that the sensations experienced in the perusal of such a document, when addressed to one's self through a public newspaper, and signed by fifteen or twenty responsible persons, are peculiar and by no means agreeable. For a moment I really began to think I was a very bad man, and that there must be something uncommonly reprehensible in my conduct.

Upon the whole, I felt that I was a little in fault, and had better apologize. There was no particular necessity for introducing Queen Victoria's front teeth and Jenny Lind's black eye to Congress; and, to confess the truth, it was really going a little beyond the usual limits of official etiquette to "ring in" a public town possessing some valuable political influence.

I therefore prepared and published in the newspapers an Apology, which it seemed to me ought to be satisfactory. The following is as close a copy of the original as I can now write out from memory:

"San Francisco, Cal., April 1st, 1858.

"To Messrs. J. Hodges, B. Punch, T. Thatcher, B. Fletcher, Warren Hastings, Wm. Pitt, J. Fox, E. Burke, and eleven others, citizens, Port Townsend, W. T.:

"Gentlemen,—I have read with surprise and regret your letter of the 10th ult., in which you make several very serious charges against me in reference to certain statements contained in my report on the Indians of Puget's Sound. Not the least important of these charges is that I stepped aside from the line of my duty to traduce your fair name and reputation as citizens of Port Townsend. You entertain the opinion that I might have been better employed — an opinion in which I would cheerfully concur if it were not based upon erroneous premises. I have not the slightest recollection of having traduced 'your fair name and reputation,' or made any reference to you whatever in my report. When I alluded to the 'beach-combers, rowdies, and other bad characters' in Port Townsend, I had no idea that respectable gentlemen like yourselves would take it as personal. Of course, as none of you ever sold whisky to the Indians or committed murder, you do great injustice to your own reputation in supposing that the public at large would attribute these crimes to you because I mentioned them in my report.

"You deny positively that either Queen Victoria or Jenny Lind had her front teeth knocked out by the Duke of York. Well, I take that back, for I certainly did not examine their mouths as closely as you seem to have done. But when you deny that Jenny Lind's eye was black, you do me great injustice. I shall insist upon it to the latest hour of my existence that it was black—deeply, darkly, beautifully black, with a prismatic circle of pink, blue, and yellow in the immediate vicinity. I cheerfully retract the teeth, but, gentlemen, I hold on to the eye. Depend upon it, I shall stand by that eye as long as the flag of freedom waves over this glorious republic! You will admit, at all events, that Jenny had a drop in her eye.

"While you do not pretend to say that there is no whisky sold in Port Townsend, you do insist upon it that I never saw any of you drunk. Of course not, gentlemen. There are several of you that I do not recollect having ever seen, either drunk or sober. If I did see any of you under the influence of intoxicating spirits, the disguise was certainly effectual, for I am now entirely unable to say which of you it was. Besides, I never said I saw any of you drunk. It requires a great deal of whisky to intoxicate some people, and I should be sorry to hazard a conjecture as to the gauge of any citizen of Port Townsend. I do not believe you habitually drink whisky as a beverage—certainly not Port Townsend whisky, for that would kill the strongest man that ever lived in less than six months, if he drank nothing else. Many of you, no doubt, use tea or coffee at breakfast, and it is quite possible that some of you occasionally venture upon water.

"Gentlemen, you were pleased to call my attention to certain custom-house claims, Indian claims, and pre-emption claims when I was at Port Townsend; but when you 'claim to be as orderly, industrious, and law-abiding as the citizens of any other town on the Pacific coast or elsewhere,' you go altogether beyond my official jurisdiction. I think you had better send that claim to Congress.

"That 'it is not customary for public officers to adhere strictly to facts for their reports' is a melancholy truth. You have me there, gentlemen. Truth is very scarce in official documents. It is not expected by the public, and it would be utterly thrown away upon Congress. Besides, the truth is the last thing that would serve your purpose as claimants for public money.

"You are charitable enough to suppose that I may have been grossly deceived in my sources of information. Well, you ought to know all about that, for I got most of the information from yourselves. As to my remark that Port Townsend is 'a benighted place,' I am astonished that you did not see into the true meaning of that expression. It was merely a jocular allusion to the absence of lamps in the public streets at night.

"You do not think it can possibly have escaped my memory that I found you engaged in your peaceful avocations as useful and respectable members of society on the occasion of my visit to Port Townsend. Now, upon my honor, I can not remember who it was particularly that I saw engaged in peaceful avocations, but I certainly saw a good many white men lying about in sunny places fast asleep, and a good many more sitting on logs of wood whittling small sticks, and apparently waiting for somebody to invite them into the nearest saloon; others I saw playing billiards, and some few standing about the corners of the streets, waiting for the houses to grow—all of which were unquestionably peaceful, if not strictly useful avocations. I have no recollection of having seen any person engaged in the performance of any labor calculated to strain his vertebræ.

"The result of your inquiries on the subject of murder appears to be that only two murders were committed in Port Townsend during the past year, instead of six, as stated in my report. Well, gentlemen, I was not present, and did not participate in any of these alleged murders, and cheerfully admit that your sheriff, who gave me the information, and whose name is appended to your letter, may not have counted them accurately. At all events, I take four of them back, and place them to the credit of Port Townsend for the ensuing year. I utterly disclaim having invented them, though I would at any time much rather invent four murders than commit one. Nor can I admit that I was at a loss for something to say. There was abundance of fictitious material presented in the course of my official investigations, without rendering it at all necessary for me to resort to imaginary murders. And I farther insist upon it that, if I did not personally witness the violent death of six men in Port Townsend, I heard the king's English most cruelly murdered there on at least six different occasions. Gentlemen, you need not take any farther trouble about 'setting yourselves right before the world.' I trust you will admit that you are all right now, since I have duly made the amende honorable.

"Wishing you success in your 'peaceful avocations,' and exemption from all future anxiety relative to the price of lots in Port Townsend, I remain, very respectfully, your obedient servant," etc.

Strange to say, so far from being satisfied with this apology, the citizens of Port Townsend were enraged to a degree bordering on insanity. The mayor, upon the reception of the mail containing the fatal document, called the Town Council together, and the schoolmaster read it to the Town Council, and the Town Council deliberated over it for three days, and then unanimously resolved that the author was a "Vile Kalumater, unworthy of further Atension, and had beter stere cleer of Port Townsend for the Future!" For two years they did nothing else, in an official point of view, but write letters to the San Francisco papers denouncing the author of this Vile Kalumy, and assuring the public that his description of Port Townsend was wholly unworthy of credit; that Port Townsend was the neatest, cleanest, most orderly, and most flourishing little town on the Pacific coast. By the time the Frazer River excitement broke out, the people of California were well acquainted, through the newspapers, with at least one town on Puget's Sound. If they knew nothing of Whatcomb, Squill-Chuck, and other rival places that aspired to popular favor, they were no strangers to the reputation of Port Townsend. Thousands, who had no particular business there, went to take a look at this wonderful town, which had given rise to so much controversy. The citizens were soon forced to build a fine hotel. Many visitors liked the society, and concluded to remain. Others thought it would soon be the great centre of commerce for all the shipping that would be drawn thither by the mineral wealth of Frazer River, and bought city lots on speculation. Traders came there and set up stores; new whisky saloons were built; customers crowded in from all parts; in short, it became a gay and dashing sort of place, and very soon had quite the appearance of a city. When the Frazer River bubble burst, nobody was killed at Port Townsend, because it had a strong reputation, and could still persuade people that it was bound to be a great city at some future period.

During the following year I made bold to pay my old friends a visit. A delegation of the Common Council met me on the wharf. There were no hacks yet introduced, but any number of horses were placed at my disposal. The greeting was cordial and impressive. A most complimentary address was read to me by the mayor of the city, in which it was fully and frankly acknowledged that I was the means of building up the fortunes of Port Townsend. After the address, the citizens with one accord rushed to me, and, grasping me warmly by the hand, at once retracted their injurious imputations. These gratifying public demonstrations over, we adjourned to the nearest saloon, and buried the hatchet forever in an ocean of the best Port Townsend whisky. It is due to the citizens to say that not one of them went beyond reasonable bounds on this joyous occasion, by which I do not mean to intimate that they were accustomed to the beverage referred to. At all events, I think it has been clearly demonstrated by these authentic documents that "whisky built a great city."

III.

THE INDIANS OF CALIFORNIA.

When the State of California was admitted into the Union, the number of Indians within its borders was estimated at one hundred thousand. Of these, some five or six thousand, residing in the vicinity of the Missions, were partially civilized, and subsisted chiefly by begging and stealing. A few of the better class contrived to avoid starvation by casual labor in the vineyards and on the farms of the settlers. They were very poor and very corrupt, given to gambling, drinking, and other vices prevailing among white men, and to which Indians have a natural inclination. As the country became more settled, it was considered profitable, owing to the high rate of compensation for white labor, to encourage these Christian tribes to adopt habits of industry, and they were employed very generally throughout the state. In the vine-growing districts they were usually paid in native brandy every Saturday night, put in jail next morning for getting drunk, and bailed out on Monday to work out the fine imposed upon them by the local authorities. This system still prevails in Los Angeles, where I have often seen a dozen of these miserable wretches carried to jail roaring drunk of a Sunday morning. The inhabitants of Los Angeles are a moral and intelligent people, and many of them disapprove of the custom on principle, and hope it will be abolished as soon as the Indians are all killed off. Practically, it is not a bad way of bettering their condition; for some of them die every week from the effects of debauchery, or kill one another in the nocturnal brawls which prevail in the outskirts of the Pueblo.

THE DIGGERS AT HOME.

The settlers in the northern portions of the state had a still more effectual method of encouraging the Indians to adopt habits of civilization. In general, they engaged them at a fixed rate of wages to cultivate the ground, and during the season of labor fed them on beans, and gave them a blanket or a shirt each; after which, when the harvest was secured, the account was considered squared, and the Indians were driven off to forage in the woods for themselves and families during the winter. Starvation usually wound up a considerable number of the old and decrepit ones every season; and of those that failed to perish from hunger or exposure, some were killed on the general principle that they must have subsisted by stealing cattle, for it was well known that cattle ranged in the vicinity, while others were not unfrequently slaughtered by their employers for helping themselves to the refuse portions of the crop which had been left in the ground. It may be said that these were exceptions to the general rule; but if ever an Indian was fully and honestly paid for his labor by a white settler, it was not my luck to hear of it; certainly it could not have been of frequent occurrence.

The wild Indians inhabiting the Coast Range, the valleys of the Sacramento and San Joaquin, and the western slope of the Sierra Nevada, became troublesome at a very early period after the discovery of the gold mines. It was found convenient to take possession of their country without recompense, rob them of their wives and children, kill them in every cowardly and barbarous manner that could be devised, and when that was impracticable, drive them as far as possible out of the way. Such treatment was not consistent with their rude ideas of justice. At best they were an ignorant race of Diggers, wholly unacquainted with our enlightened institutions. They could not understand why they should be murdered, robbed, and hunted down in this way, without any other pretense of provocation than the color of their skin and the habits of life to which they had always been accustomed. In the traditionary researches of their most learned sages they had never heard of the snakes in Ireland that were exterminated for the public benefit by the great and good St. Patrick. They were utterly ignorant of the sublime doctrine of General Welfare. The idea, strange as it may appear, never occurred to them that they were suffering for the great cause of civilization, which, in the natural course of things, must exterminate Indians. Actuated by base motives of resentment, a few of them occasionally rallied, preferring rather to die than submit to these imaginary wrongs. White men were killed from time to time; cattle were driven off; horses were stolen, and various other iniquitous offenses were committed.

The federal government, as is usual in cases where the lives of valuable voters are at stake, was forced to interfere. Troops were sent out to aid the settlers in slaughtering the Indians. By means of mounted howitzers, muskets, Minié rifles, dragoon pistols, and sabres, a good many were cut to pieces. But, on the whole, the general policy of the government was pacific. It was not designed to kill any more Indians than might be necessary to secure the adhesion of the honest yeomanry of the state, and thus furnish an example of the practical working of our political system to the savages of the forest, by which it was hoped they might profit. Congress took the matter in hand at an early day, and appropriated large sums of money for the purchase of cattle and agricultural implements. From the wording of the law, it would appear that these useful articles were designed for the relief and maintenance of the Indians. Commissioners were appointed at handsome salaries to treat with them, and sub agents employed to superintend the distribution of the purchases. In virtue of this munificent policy, treaties were made in which the various tribes were promised a great many valuable presents, which of course they never got. There was no reason to suppose they ever should; it being a fixed principle with strong powers never to ratify treaties made by their own agents with weaker ones, when there is money to pay and nothing to be had in return.

The cattle were purchased, however, to the number of many thousands. Here arose another difficulty. The honest miners must have something to eat, and what could they have more nourishing than fat cattle? Good beef has been a favorite article of subsistence with men of bone and muscle ever since the days of the ancient Romans. So the cattle, or the greater part of them, were driven up to the mines, and sold at satisfactory rates—probably for the benefit of the Indians, though I never could understand in what way their necessities were relieved by this speculation, unless it might be that the parties interested turned over to them the funds received for the cattle. It is very certain they continued to starve and commit depredations in the most ungrateful manner for some time after; and, indeed, to such a pitch of audacity did they carry their rebellious spirit against the constituted authorities, that many of the chiefs protested if the white people would only let them alone, and give them the least possible chance to make a living, they would esteem it a much greater favor than any relief they had experienced from the munificent donations of Congress.

But government was not to be defeated in its benevolent intentions. Voluminous reports were made to Congress, showing that a general reservation system, on the plan so successfully pursued by the Spanish missionaries, would best accomplish the object. It was known that the Missions of California had been built chiefly by Indian labor; that during their existence the priests had fully demonstrated the capacity of this race for the acquisition of civilized habits; that extensive vineyards and large tracts of land had been cultivated solely by Indian labor, under their instruction; and that by this humane system of teaching many hostile tribes had been subdued, and enabled not only to support themselves, but to render the Missions highly profitable establishments.

No aid was given by government beyond the grants of land necessary for missionary purposes; yet they soon grew wealthy, owned immense herds of cattle, supplied agricultural products to the rancheros, and carried on a considerable trade in hides and tallow with the United States. If the Spanish priests could do this without arms or assistance, in the midst of a savage country, at a period when the Indians were more numerous and more powerful than they are now, surely it could be done in a comparatively civilized country by intelligent Americans, with all the lights of experience and the co-operation of a beneficent government.

At least Congress thought so; and in 1853 laws were passed for the establishment of a reservation system in California, and large appropriations were made to carry it into effect. Tracts of land of twenty-five thousand acres were ordered to be set apart for the use of the Indians; officers were appointed to supervise the affairs of the service; clothing, cattle, seeds, and agricultural implements were purchased; and a general invitation was extended to the various tribes to come in and learn how to work like white men. The first reservation was established at the Tejon, a beautiful and fertile valley in the southern part of the state. Head-quarters for the employés, and large granaries for the crops, were erected. The Indians were feasted on cattle, and every thing promised favorably. True, it cost a great deal to get started, about $250,000; but a considerable crop was raised, and there was every reason to hope that the experiment would prove successful. In the course of time other reservations were established, one in the foot-hills of the Sacramento Valley, at a place called Nome Lackee; one at the mouth of the Noyo River, south of Cape Mendocino; and one on the Klamath, below Crescent City; besides which, there were Indian farms, or adjuncts, of these reservations at the Fresno, Nome Cult or Round Valley, the Mattole Valley, near Cape Mendocino, and other points where it was deemed advisable to give aid and instruction to the Indians. The cost of these establishments was such as to justify the most sanguine anticipations of their success.

In order that the appropriations might be devoted to their legitimate purpose, and the greatest possible amount of instruction furnished at the least expense, the Executive Department adopted the policy of selecting officers experienced in the art of public speaking, and thoroughly acquainted with the prevailing systems of primary elections. A similar policy had been found to operate beneficially in the case of Collectors of Customs, and there was no reason why it should not in other branches of the public service. Gentlemen skilled in the tactics of state Legislatures, and capable of influencing those refractory bodies by the exercise of moral suasion, could be relied upon to deal with the Indians, who are not so far advanced in the arts of civilization, and whose necessities, in a pecuniary point of view, are not usually so urgent. Besides, it was known that the Digger tribes were exceedingly ignorant of our political institutions, and required more instruction, perhaps, in this branch of knowledge than in any other. The most intelligent of the chiefs actually had no more idea of the respective merits of the great candidates for senatorial honors in California than if those distinguished gentlemen had never been born. As to primary meetings and caucuses, the poor Diggers, in their simplicity, were just as apt to mistake them for some favorite game of thimblerig or pitch-penny as for the practical exercise of the great system of free suffrage. They could not make out why men should drink so much whisky and swear so hard unless they were gambling; and if any farther proof was necessary, it was plain to see that the game was one of hazard, because the players were constantly whispering to each other, and passing money from hand to hand, and from pocket to pocket. The only difference they could see between the different parties was that some had more money than others, but they had no idea where it came from. To enlighten them on all these points was, doubtless, the object of the great appointing powers in selecting good political speakers to preside over them. After building their houses, it was presumed that there would be plenty of stumps left in the woods from which they could be taught to make speeches on the great questions of the day, and where a gratifying scene might be witnessed, at no remote period, of big and little Diggers holding forth from every stump in support of the presiding administration. For men who possessed an extraordinary capacity for drinking ardent spirits; who could number among their select friends the most notorious vagrants and gamblers in the state; who spent their days in idleness and their nights in brawling grog-shops; whose habits, in short, were in every way disreputable, the authorities in Washington entertained a very profound antipathy. I know this to be the case, because the most stringent regulations were established prohibiting persons in the service from getting drunk, and official orders written warning them that they would be promptly removed in case of any misconduct. Circular letters were also issued, and posted up at the different reservations, forbidding the employés to adopt the wives of the Indians, which it was supposed they might attempt to do from too zealous a disposition to cultivate friendly relations with both sexes. In support of this policy, the California delegation made it a point never to indorse any person for office in the service who was not considered peculiarly deserving of patronage. They knew exactly the kind of men that were wanted, because they lived in the state and had read about the Indians in the newspapers. Some of them had even visited a few of the wigwams. Having the public welfare at heart—a fact that can not be doubted, since they repeatedly asserted it in their speeches—they saw where the great difficulty lay, and did all in their power to aid the executive. They indorsed the very best friends they had—gentlemen who had contributed to their election, and fought for them through thick and thin. The capacity of such persons for conducting the affairs of a reservation could not be doubted. If they had cultivated an extensive acquaintance among pot-house voters, of course they must understand the cultivation of potatoes and onions; if they could control half a dozen members of the Legislature in a senatorial contest, why not be able to control Indians, who were not near so difficult to manage? if they could swallow obnoxious measures of the administration, were they not qualified to teach savages how to swallow government provisions? if they were honest enough to avow, in the face of corrupt and hostile factions, that they stood by the Constitution, and always meant to stand by the same broad platform, were they not honest enough to disburse public funds?

In one respect, I think the policy of the government was unfortunate—that is, in the disfavor with which persons of intemperate and disreputable habits were regarded. Men of this kind—and they are not difficult to find in California—could do a great deal toward meliorating the moral condition of the Indians by drinking up all the whisky that might be smuggled on the reservations, and behaving so disreputably in general that no Indian, however degraded in his propensities, could fail to become ashamed of such low vices.

In accordance with the views of the Department, it was deemed to be consistent with decency that these untutored savages should be clothed in a more becoming costume than Nature had bestowed upon them. Most of them were as ignorant of covering as they were of the Lecompton Constitution. With the exception of a few who had worked for the settlers, they made their first appearance on the reservations very much as they appeared when they first saw daylight. It was a great object to make them sensible of the advantages of civilization by covering their backs while cultivating their brains. Blankets, shirts, and pantaloons, therefore, were purchased for them in large quantities. It is presumed that when the Department read the vouchers for these articles, and for the potatoes, beans, and cattle that were so plentifully sprinkled through the accounts, it imagined that it was "clothing the naked and feeding the hungry!"

The blankets, to be sure, were very thin, and cost a great deal of money in proportion to their value; but, then, peculiar advantages were to be derived from the transparency of the fabric. In some respects the worst material might be considered the most economical. By holding his blanket to the light, an Indian could enjoy the contemplation of both sides of it at the same time; and it would only require a little instruction in architecture to enable him to use it occasionally as a window to his wigwam. Every blanket being marked by a number of blotches, he could carry his window on his back whenever he went out on a foraging expedition, so as to know the number of his residence when he returned, as the citizens of Schilda carried their doors when they went away from home, in order that they should not forget where they lived. Nor was it the least important consideration, that when he gambled it away, or sold it for whisky, he would not be subject to any inconvenience from a change of temperature. The shirts and pantaloons were in general equally transparent, and possessed this additional advantage, that they very soon cracked open in the seams, and thereby enabled the squaws to learn how to sew.

As many of the poor wretches were afflicted with diseases incident to their mode of life, and likely to contract others from the white employés of the reservations, physicians were appointed to give them medicine. Of course Indians required a peculiar mode of treatment. They spoke a barbarous jargon, and it was not possible that any thing but barbarous compounds could operate on their bowels. Of what use would it be to waste good medicines on stomachs that were incapable of comprehending their use? Accordingly, any deficiency in the quality was made up by the quantity and variety. Old drug stores were cleared of their rubbish, and vast quantities of croton oil, saltpetre, alum, paint, scent-bottles, mustard, vinegar, and other valuable laxatives, diaphoretics, and condiments were supplied for their use. The result was, that, aided by the peculiar system of diet adopted, the physicians were enabled very soon to show a considerable roll of patients. In cases where the blood was ascertained to be scorbutic, the patients were allowed to go out in the valleys, and subsist for a few months on clover or grass, which was regarded as a sovereign remedy. I was assured at one reservation that fresh spring grass had a more beneficial effect on them than the medicines, as it generally purged them. The Department was fully advised of these facts in elaborate reports made by its special emissaries, and congratulated itself upon the satisfactory progress of the system. The elections were going all right—the country was safe. Feeding Indians on grass was advancing them at least one step toward a knowledge of the sacred Scriptures. It was following the time-honored precedent of Nebuchadnezzar, the King of Babylon, who was driven from men, and did eat grass as oxen, and was wet with the dews of heaven till his hairs were grown like eagles' feathers, and his nails like birds' claws. An ounce of croton oil would go a great way in lubricating the intestines of an entire tribe of Indians; and if the paint could not be strictly classed with any of the medicines known in the official dispensary, it might at least be used for purposes of clothing during the summer months. Red or green pantaloons painted on the legs of the Indians, and striped blue shirts artistically marked out on their bodies, would be at once cool, economical, and picturesque. If these things cost a great deal of money, as appeared by the vouchers, it was a consolation to know that, money being the root of all evil, no injurious effects could grow out of such a root after it had been once thoroughly eradicated.

The Indians were also taught the advantages to be derived from the cultivation of the earth. Large supplies of potatoes were purchased in San Francisco, at about double what they were worth in the vicinity of the reservations. There were only twenty-five thousand acres of public land available at each place for the growth of potatoes or any other esculent for which the hungry natives might have a preference; but it was much easier to purchase potatoes than to make farmers out of the white men employed to teach them how to cultivate the earth. Sixteen or seventeen men on each reservation had about as much as they could do to attend to their own private claims, and keep the natives from eating their private crops. It was not the policy of government to reward its friends for their "adhesion to the Constitution" by requiring them to perform any practical labor at seventy-five or a hundred dollars a month, which was scarcely double the current wages of the day. Good men could obtain employment any where by working for their wages; but it required the best kind of administration men to earn extraordinary compensations by an extraordinary amount of idleness. Not that they were all absolutely worthless. On the contrary, some spent their time in hunting, others in riding about the country, and a considerable number in laying out and supervising private claims, aided by Indian labor and government provisions.

The official reports transmitted to Congress from time to time gave flattering accounts of the progress of the system. The extent and variety of the crops were fabulously grand. Immense numbers of Indians were fed and clothed—on paper. Like little children who cry for medicines, it would appear that the whole red race were so charmed with the new schools of industry that they were weeping to be removed there and set to work. Indeed, many of them had already learned to work "like white men;" they were bending to it cheerfully, and could handle the plow and the sickle very skillfully, casting away their bows and arrows, and adopting the more effective instruments of agriculture. No mention was made of the fact that these working Indians had acquired their knowledge from the settlers, and that, if they worked after the fashion of the white men on the reservations, it was rarely any of them were obliged to go to the hospital in consequence of injuries resulting to the spinal column. The favorite prediction of the officers in charge was, that in a very short time these institutions would be self-sustaining—that is to say, that neither they nor the Indians would want any more money after a while.

It may seem strange that the appropriations demanded of Congress did not decrease in a ratio commensurate with these flattering reports. The self-sustaining period had not yet come. On the contrary, as the Indians were advancing into the higher branches of education —music, dancing, and the fine arts, moral philosophy and ethics, political economy, etc.—it required more money to teach them. The number had been considerably diminished by death and desertion; but then their appetites had improved, and they were getting a great deal smarter. Besides, politics were becoming sadly entangled in the state, and many agents had to be employed in the principal cities to protect the women and children from any sudden invasion of the natives while the patriotic male citizens were at the polls depositing their votes.

The Department, no doubt, esteemed all this to be a close approximation to the Spanish Mission system, and in some respects it was. The priests sought the conversion of heathens, who believed neither in the Divinity nor the Holy Ghost; the Department the conversion of infidels, who had no faith in the measures of the administration. If there was any material difference, it was in the Head of the Church, and the missionaries appointed to carry its views into effect.

But the most extraordinary feature in the history of this service in California was the interpretation given by the federal authorities in Washington to the Independent Treasury Act of 1846. That stringent provision, prohibiting any public officer from using for private purposes, loaning, or depositing in any bank or banking institution any public funds committed to his charge; transmitting for settlement any voucher for a greater amount than that actually paid; or appropriating such funds to any other purpose than that prescribed by law, was so amended in the construction of the Department as to mean, "except in cases where such officer has rendered peculiar services to the party and possesses strong influences in Congress." When any infraction of the law was reported, it was subjected to the test of this amended reading; and if the conditions were found satisfactory, the matter was disposed of in a pigeon-hole. An adroit system of accountability was established, by which no property return, abstract of issues, account current, or voucher, was understood to mean what it expressed upon its face, so that no accounting officer possessing a clew to the policy adopted could be deceived by the figures. Thus it was perfectly well understood that five hundred or a thousand head of cattle did not necessarily mean real cattle with horns, legs, and tails, actually born in the usual course of nature, purchased for money, and delivered on the reservations, but prospective cattle, that might come into existence and be wanted at some future period. For all the good the Indians got of them, it might as well be five hundred or a thousand head of voters, for they no more fed upon beef, as a general thing, than they did upon human flesh.

Neither was it beyond the capacity of the Department to comprehend that traveling expenses on special Indian service might just as well mean a trip to the Convention at Sacramento; that guides and assistants were a very indefinite class of gentlemen of a roving turn of mind; that expenses incurred in visiting wild tribes and settling difficulties among them did not necessarily involve the exclusion of difficulties among the party factions in the Legislature. In short, the original purpose of language was so perverted in the official correspondence that it had no more to do with the expression of facts than many of the employés had to do with the Indians. The reports and regulations of the Department actually bordered on the poetical. It was enough to bring tears into the eyes of any feeling man to read the affecting dissertations that were transmitted to Congress on the woes of the Red men, and the labors of the public functionaries to meliorate their unhappy condition. Faith, hope, and charity abounded in them. "See what we are doing for these poor children of the forest!" was the burden of the song, in a strain worthy the most pathetic flights of Mr. Pecksniff; "see how faithful we are to our trusts, and how judiciously we expend the appropriations! Yet they die off in spite of us—wither away as the leaves of the trees in autumn! Let us hope, nevertheless, that the beneficent intentions of Congress may yet be realized. We are the guardians of these unfortunate and defenseless beings; they are our wards; it is our duty to take care of them; we can afford to be liberal, and spend a little more money on them. Through the judicious efforts of our public functionaries, and the moral influences spread around them, there is reason to believe they will yet embrace civilization and Christianity, and become useful members of society." In accordance with these views, the regulations issued by the Department were of the most stringent character—encouraging economy, industry, and fidelity; holding all agents and employés to a strict accountability; with here and there some instructive maxim of morality—all of which, upon being translated, meant that politicians are very smart fellows, and it was not possible for them to humbug one another. "Do your duty to the Indians as far as you can conveniently, and without too great a sacrifice of money; but stand by our friends, and save the party by all means and at all hazards. *Verbum sap!*" was the practical construction.

When public clamor called attention to these supposed abuses, and it became necessary to make some effective demonstration of honesty, a special agent was directed to examine into the affairs of the service and report the result. It was particularly enjoined upon him to investigate every complaint affecting the integrity of public officers, collect and transmit the proofs of malfeasance, with his own views in the premises, so that every abuse might be uprooted and cast out of the service. Decency in official conduct must be respected and the public eye regarded! Peremptory measures would be taken to suppress all frauds upon the Treasury. It was the sincere desire of the administration to preserve purity and integrity in the public service.

From mail to mail, during a period of three years, the agent made his reports; piling up proof upon proof, and covering acres of valuable paper with protests and remonstrances against the policy pursued; racking his brains to do his duty faithfully; subjecting himself to newspaper abuse for neglecting it, because no beneficial result was perceptible, and making enemies as a matter of course. Reader, if ever you aspire to official honors, let the fate of that unfortunate agent be a warning to you. He did exactly what he was instructed to do, which was exactly what he was not wanted to do. In order to save time and expense, as well as farther loss of money in the various branches of public service upon which he had reported, other agents were sent out to ascertain if he had told the truth; and when they were forced to admit that he had, there was a good deal of trouble in the wigwam of the great chief. Not only did poor Yorick incur the hostility of powerful senatorial influences, but by persevering in his error, and insisting that he had told the truth, the whole truth, and nothing but the truth, he eventually lost the respect and confidence of the "powers that be," together with his official head. I knew him well. He was a fellow of infinite jest. There was something so exquisitely comic in the idea of taking official instructions literally, and carrying them into effect, that he could not resist it. The humor of the thing kept him in a constant chuckle of internal satisfaction; but it was the most serious jest he ever perpetrated, for it cost him, besides the trouble of carrying it out, the loss of a very comfortable per diem.

The results of the policy pursued were precisely such as might have been expected. A very large amount of money was annually expended in feeding white men and starving Indians. Such of the latter as were physically able took advantage of the tickets-of-leave granted them so freely, and left. Very few ever remained at these benevolent institutions when there was a possibility of getting any thing to eat in the woods. Every year numbers of them perished from neglect and disease, and some from absolute starvation. When it was represented in the official reports that two or three thousand enjoyed the benefit of aid from government within the limits of each district—conveying the idea that they were fed and clothed at public expense—it must have meant that the Territory of California originally cost the United States fifteen millions of dollars, and that the nuts and berries upon which the Indians subsisted, and the fig-leaves in which they were supposed to be clothed, were embraced within the cessions made by Mexico. At all events, it invariably happened, when a visitor appeared on the reservations, that the Indians were "out in the mountains gathering nuts and berries." This was the case in spring, summer, autumn, and winter. They certainly possessed a remarkable predilection for staying out a long time. Very few of them, indeed, have yet come back. The only difference between the existing state of things and that which existed prior to the inauguration of the system is, that there were then some thousands of Indians living within the limits of the districts set apart for reservation purposes, whereas there are now only some

OUT IN THE MOUNTAINS.

hundreds. In the brief period of six years they have been very nearly destroyed by the generosity of government. What neglect, starvation, and disease have not done, has been achieved by the co-operation of the white settlers in the great work of extermination.

No pretext has been wanted, no opportunity lost, whenever it has been deemed necessary to get them out of the way. At Nome Cult Valley, during the winter of 1858–'59, more than a hundred and fifty peaceable Indians, including women and children, were cruelly slaughtered by the whites who had settled there under official authority, and most of whom derived their support either from actual or indirect connection with the reservation. Many of them had been in public employ, and now enjoyed the rewards of their meritorious services. True, a notice was posted up on the trees that the valley was public land reserved for Indian purposes, and not open to settlement; but nobody, either in or out of the service, paid any attention to that, as a matter of course. When the Indians were informed that it was their home, and were invited there on the pretext that they would be protected, it was very well understood that, as soon as government had spent money enough there to build up a settlement sufficiently strong to maintain itself, they would enjoy very slender chances of protection. It was alleged that they had driven off and eaten private cattle. There were some three or four hundred head of public cattle on the property returns, all supposed to be ranging in the same vicinity; but the private cattle must have been a great deal better, owing to some superior capacity for eating grass. Upon an investigation of this charge, made by the officers of the army, it was found to be entirely destitute of truth: a few cattle had been lost, or probably killed by white men, and this was the whole basis of the massacre. Armed parties went into the rancherias in open day, when no evil was apprehended, and shot the Indians down—weak, harmless, and defenseless as they were — without distinction of age or sex;

shot down women with sucking babes at their breasts; killed or crippled the naked children that were running about; and, after they had achieved this brave exploit, appealed to the state government for aid! Oh, Shame, Shame, where is thy blush, that white men should do this with impunity in a civilized country, under the very eyes of an enlightened government! They did it, and they did more! For days, weeks, and months they ranged the hills of Nome Cult, killing every Indian that was too weak to escape; and, what is worse, they did it under a state commission, which in all charity I must believe was issued upon false representations. A more cruel series of outrages than those perpetrated upon the poor Indians of Nome Cult never disgraced a community of white men. The state said the settlers must be protected, and it protected them—protected them from women and children, for the men are too imbecile and too abject to fight. The general government folded its arms and said, "What can we do? We can not chastise the citizens of a state. Are we not feeding and clothing the savages, and teaching them to be moral, and is not that as much as the civilized world can ask of us?"

At King's River, where there was a public farm maintained at considerable expense, the Indians were collected in a body of two or three hundred, and the white settlers, who complained that government would not do any thing for them, drove them over to the Agency at the Fresno. After an expenditure of some thirty thousand dollars a year for six years, that farm had scarcely produced six blades of grass, and was entirely unable to support over a few dozen Indians who had always lived there, and who generally foraged for their own subsistence. The new-comers, therefore, stood a poor chance till the agent purchased from the white settlers, on public account, the acorns which they (the Indians) had gathered and laid up for winter use at King's River. Notwithstanding the acorns, they were very soon starved out at the Fresno, and wandered away to find a subsistence wherever they could. Many of them perished of hunger on the plains of the San Joaquin. The rest are presumed to be in the mountains gathering berries.

At the Mattole Station, near Cape Mendocino, a number of Indians were murdered on the public farm within a few hundred yards of the head-quarters. The settlers in the valley alleged that government would not support them, or take any care of them; and as settlers were not paid for doing it, they must kill them to get rid of them.

At Humboldt Bay, and in the vicinity, a series of Indian massacres by white men continued for over two years. The citizens held public meetings, and protested against the action of the general government in leaving these Indians to prowl upon them for a support. It was alleged that the reservations cost two hundred and fifty thousand dollars a year, and yet nothing was done to relieve the people of this burden. Petitions were finally sent to the state authorities asking for the removal of the Indians from that vicinity; and the state sent out its militia, killed a good many, and captured a good many others, who were finally carried down to the Mendocino reservation. They liked that place so well that they left it very soon, and went back to their old places of resort, preferring a chance of life to the certainty of starvation. During the winter of last year a number of them were gathered at Humboldt. The whites thought it was a favorable opportunity to get rid of them altogether. So they went in a body to the Indian camp, during the night when the poor wretches were asleep, shot all the men, women, and children they could at the first onslaught, and cut the throats of the remainder. Very few escaped. Next morning sixty bodies lay weltering in their blood—the old and the young, male and female—with every wound gaping a tale of horror to the civilized world. Children climbed upon their mothers' breasts, and sought nourishment from the fountains that death had drained; girls and boys lay here and there with their throats cut from ear to ear; men and women, cling-

PROTECTING THE SETTLERS.

ing to each other in their terror, were found perforated with bullets or cut to pieces with knives—all were cruelly murdered! Let any one who doubts this read the newspapers of San Francisco of that date. It will be found there in its most bloody and tragic details. Let them read of the Pitt River massacre, and of all the massacres that for the past three years have darkened the records of the state.

I will do the white people who were engaged in these massacres the justice to say that they were not so much to blame as the general government. They had at least given due warning of their intention. For years they had burdened the mails with complaints of the inefficiency of the agents; they had protested in the newspapers, in public meetings, in every conceivable way, and on every possible occasion, against the impolicy of permitting these Indians to roam about the settlements, picking up a subsistence in whatever way they could, when there was a fund of $250,000 a year appropriated by Congress for their removal to and support on the reservations. What were these establishments for? Why did they not take charge of the Indians? Where were the agents? What was done with the money? It was repeatedly represented that, unless something was done, the Indians would soon all be killed. They could no longer make a subsistence in their old haunts. The progress of settlement had driven them from place to place till there was no longer a spot on earth they could call their own. Their next move could only be into the Pacific Ocean. If ever an unfortunate people needed a few acres of ground to stand upon, and the poor privilege of making a living for themselves, it was these hapless Diggers. As often as they tried the reservations, sad experience taught them that these were institutions for the benefit of white men, not Indians. It was wonderful how the employés had prospered on their salaries. They owned fine ranches in the vicinity; in fact, the reservations themselves were pretty much covered with the claims of persons in the service, who thought they would make nice farms for white men. The principal work done was to attend to sheep and cattle speculations, and make shepherds out of the few Indians that were left.

What did it signify that thirty thousand dollars a year had been expended at the Tejon? thirty thousand at the Fresno? fifty thousand at Nome Lackee? ten thousand at Nome Cult? forty-eight thousand at Mendocino? sixteen thousand at the Klamath? and some fifty or sixty thousand for miscellaneous purposes? that all this had resulted in the reduction of a hundred thousand Indians to about thirty thousand? Meritorious services had been rewarded, and a premium in favor of public integrity issued to an admiring world.

I am satisfied, from an acquaintance of eleven years with the Indians of California, that, had the least care been taken of them, these disgraceful massacres would never have occurred. A more inoffensive and harmless race of beings does not exist on the face of the earth; but, wherever they attempted to procure a subsistence, they were hunted down; driven from the reservations by the instinct of self-preservation; shot down by the settlers upon the most frivolous pretexts; and abandoned to their fate by the only power that could have afforded them protection.

This was the result, in plain terms, of the inefficient and discreditable manner in which public affairs were administered by the federal authorities in Washington. It was the natural consequence of a corrupt political system, which, for the credit of humanity, it is to be hoped will be abandoned in future so far as the Indians are concerned. They have no voice in public affairs. So long as they are permitted to exist, party discipline is a matter of very little moment to them. All they ask is the privilege of breathing the air that God gave to us all, and living in peace wherever it may be convenient to remove them. Their history in California is a melancholy record of neglect and cruelty; and the part taken by public men high in position, in wresting from them the very means of subsistence, is one of which any other than professional politicians would be ashamed. For the Executive Department there is no excuse. There lay the power and the remedy; but a paltry and servile spirit, an abject submission to every shifting influence, an utter absence of that high moral tone which is the characteristic trait of genuine statesmen and patriots, have been the distinguishing features of this branch of our government for some time past. Disgusted with their own handiwork; involved in debt throughout the state, after wasting all the money appropriated by Congress; the accounts in an inextricable state of confusion; the creditors of the government clamoring to be paid; the "honest yeomanry" turning against the party in power; political affairs entangled beyond remedy; it was admitted to be a very bad business—not at all such as to meet the approval of the administration. The appropriation was cut down to fifty thousand dollars. That would do damage enough. Two hundred and fifty thousand a year, for six or seven years, had inflicted sufficient injury upon the poor Indians. Now it was time to let them alone on fifty thousand, or turn them over to the state. So the end of it is, that the reservations are practically abandoned; the remainder of the Indians are being exterminated every day; and the Spanish Mission System has signally failed.

A PEEP AT WASHOE.

CHAPTER I.

When I inform the reader that I have scarcely dipped pen in ink for six years save to unravel the mysteries of a Treasury voucher; that I have lived chiefly among Indians, disbursing agents, and officers of the customs; that I now sit writing in the attic of a German villa more than eight thousand miles from the scene of my adventures, without note or memorandum of any kind to refresh my memory, he will be prepared to make reasonable allowance for such a loose, rambling, and disjointed narrative as an ex-inspector general can be expected to write under such adverse circumstances. If there be inconveniences in being hanged, as the gentle Elia has attempted to prove, so likewise are there inconveniences in being decapitated; for surely a man deprived of the casket which nature has given him as a receptacle for his brains is no better off than one with a broken neck. But it is not my present purpose to enter into an analysis of this portion of my experience; nor do I make these references to official life by way of excuse for any rustiness of intellect that may be perceptible in my narrative, but rather in mitigation of those unconscious violations of truth and marvelous flights of fancy which may naturally result from long experience in government affairs.

Ever since 1849, when I first trod the shores of California, the citizens of that Land of Promise have been subject to periodical excitements, the extent and variety of which can find no parallel in any other state of the

Union. To enumerate these in chronological detail would be a difficult task, nor is it necessary to my purpose. The destruction of towns by flood and fire; the uprisings and downfallings of vigilance committees; the breaking of banking-houses and pecuniary ruin of thousands; the political wars, senatorial tournaments, duels, and personal affrays; the prison and bulkhead schemes; the extraordinary ovations to the living and the dead, and innumerable other excitements, have been too frequently detailed, and have elicited too much comment from the Atlantic press not to be still in the memory of the public.

But, numerous as these agitations have been, and prejudicial as some of them must long continue to be to the reputation of the state, they can bear no comparison in point of extent and general interest to the mining excitements which from time to time have convulsed the whole Pacific coast, from Puget's Sound to San Diego. In these there can be no occasion for party animosity; they are confined to no political or sectional clique; all the industrial classes are interested, and in a manner, too, affecting, either directly or incidentally, their very means of subsistence. The country abounds in mineral wealth, and the merchant, the banker, the shipper, the mechanic, the laborer, are all, to some extent, dependent upon its development. Even the gentleman of elegant leisure, vulgarly known as the "Bummer"—and there are many in California—is occasionally driven by visions of cocktail and cigar-money to doff his "stove-pipe," and exchange his gold-mounted cane for a pick or a shovel. The axiom has been well established by an eminent English writer that "every man wants a thousand pounds." It seems, indeed, to be a chronic and constitutional want, as well in California as in less favored countries.

Few of the early residents of the state can have forgotten the Gold Bluff excitement of '52, when, by all accounts, old Ocean himself turned miner, and washed up cartloads of gold on the beach above Trinidad. It was

THE BUMMER.

represented, and generally believed, that any enterprising man could take his hat and a wheelbarrow, and in half an hour gather up gold enough to last him for life. I have reason to suspect that, of the thousands who went there, many will long remember their experience with emotions, if pleasant, "yet mournful to the soul."

The Kern River excitement threatened for a time to depopulate the northern portion of the state. The stages from Marysville and Sacramento were crowded day after day, and new lines were established from Los Angeles,

GOING TO KERN RIVER.

Stockton, San José, and various other points; but such was the pressure of travel in search of this grand depository, in which it was represented the main wealth of the world had been treasured by a beneficent Providence, that thousands were compelled to go on foot, and carry their blankets and provisions on their backs. From Stockton to the mining district, a distance of more than three hundred miles, the plains of the San Joaquin were literally speckled with "honest miners." It is a notable

RETURNING FROM KERN RIVER.

fact, that of those who went in stages, the majority returned on foot; and of those who trusted originally to shoe-leather, many had to walk back on their natural soles, or depend on sackcloth or charity.

After the Kern River Exchequer had been exhausted, the public were congratulated by the press throughout the state upon the effectual check now put upon these ruinous and extravagant excitements. The enterprising miners who had been tempted to abandon good claims in search of better had undergone a species of purging which would allay any irritation of the mucous membrane for some time. What they had lost in money they had gained in experience. They would henceforth turn a deaf ear to interested representations, and not be dazzled by visions of sudden wealth conjured up by monte-dealers, travelers, and horse-jockeys. They were, on the whole, wiser if not happier men. Nor would the lesson be lost to the merchants and capitalists who had scattered their goods and their funds over the picturesque heights of the Sierra Nevada. And even the gentlemen of elegant leisure, who had gone off so suddenly in search of small change for liquors and cigars, could now recuperate their exhausted energies at the free lunch establishments of San Francisco, or, if too far gone in seed for that, they could regenerate their muscular system by some wholesome exercise in the old diggings, where there was not so much gold perhaps as at Kern River, but where it could be got at more easily.

Scarcely had the reverberation caused by the bursting of the Kern River bubble died away, and fortune again smiled upon the ruined multitudes, when a faint cry was heard from afar—first low and uncertain, like a mysterious whisper, then full and sonorous, like the boom of glad tidings from the mouth of a cannon, the inspiring cry of FRAZER RIVER! Here was gold sure enough! a river of gold! a country that dazzled the eyes with its glitter of gold! There was no deception about it this time. New Caledonia was the land of Ophir. True, it was in the British possessions, but what of that? The people of California would develop the British posses-

HO! FOR FRAZER RIVER.

sions. Had our claim to 54° 40′ been insisted upon, this immense treasure would now have been within our own boundaries; but no matter — it was ours by right of proximity. The problem of Solomon's Temple was now solved. Travelers, from Marco Polo down to the present era, who had attempted to find the true land of Ophir, had signally failed; but here it was, the exact locality, beyond peradventure. For where else in the world could the river-beds, creeks, and cañons be lined with gold? Where else could the honest miner "pan out" $100 per day every day in the year? But if any who had been rendered incredulous by former excitements still doubted, they could no longer discredit the statements that were brought down by every steamer, accompanied by positive and palpable specimens of the ore, and by the assurances of captains, pursers, mates, cooks, and waiters, that Frazer River was the country. To be sure, it was afterward hinted that the best part of the gold brought down from Frazer had made the round voyage from San Francisco; but I consider this a gross and unwarranted imputation upon the integrity of steam-boat owners, captains, and speculators. Did not the famous Commodore Wright take the matter in hand; put his best steamers on the route; hoist his banners and placards in every direction, and give every man a chance of testing the question in person? This was establishing the existence of immense mineral wealth in that region upon a firm and practical basis. No man of judgment and experience, like the commodore, would undertake to run his steamers on "the baseless fabric of a vision." The cheapness and variety of his rates afforded every man an opportunity of making a fortune. For thirty, twenty, and even fifteen dollars, the ambitious aspirant for Frazer could be landed at Victoria.

I will not now undertake to give a detail of that memorable excitement; how the stages, north, south, east, and, I had almost said, west, were crowded day and night with scores upon scores of sturdy adventurers;

how farms were abandoned and crops lost for want of hands to work them; how rich claims in the old diggings were given away for a song; how the wharves of San Francisco groaned under the pressure of the human freight delivered upon them on every arrival of the Sacramento and Stockton boats; how it was often impracticable to get through the streets in that vicinity owing to the crowds gathered around the "runners," who cried aloud the merits and demerits of the rival steamers; and, strangest of all, how the head and front of the Frazerites were the very men who had enjoyed such pleasant experience at Gold Bluff, Kern River, and other places famous in the history of California. No sensible man could doubt the richness of Frazer River when these veterans became leaders, and called upon the masses to follow. They were not a class of men likely to be deceived —they knew the signs of the times. And, in addition to all this, who could resist the judgment and experience of Commodore Wright, a man who had made an independent fortune in the steam-boat business? Who could be deaf when assayers, bankers, jobbers, and speculators cried aloud that it was all true?

Well, I am not going to moralize. Mr. Nugent was appointed a commissioner, on the part of the United States, to settle the various difficulties which had grown up between the miners and Governor Douglass. He arrived at Victoria in time to perform signal service to his fellow-citizens; that is to say, he found many of them in a state of starvation, and sent them back to California at public expense. Frazer River, always too high for mining purposes, could not be prevailed upon to subside. Its banks were not banks of issue, nor were its beds stuffed with the feathers of the Golden Goose. Had it not been for this turn of affairs, it is difficult to say what would have been the result. The British Lion had been slumbering undisturbed at Victoria for half a century, and was very much astonished, upon waking up, to find thirty thousand semi-barbarous Californians scattered broadcast over the British possessions. Governor Douglass issued manifestoes in vain. He evidently thought it no joke. The subject eventually became a matter of diplomatic correspondence, in which much ink was shed, but fortunately no blood, although the subsequent seizure of San Juan by General Harney came very near producing that result.

RETURNED FROM FRAZER RIVER.

The steamers, in due course of time, began to return crowded with enterprising miners, who still believed there was gold there if the river would only fall. But generosity dictates that I should say no more on this point. It is enough to add, that the time arrived when it became a matter of personal offense to ask any spirited gentleman if he had been to Frazer River.

There was now, of course, an end to all mining excitements. It could never again happen that such an imposition could be practiced upon public credulity. In the whole state there was not another sheep that could be gulled by the cry of wolf. Business would now resume its steady and legitimate course. Property would cease to fluctuate in value. Every branch of industry would become fixed upon a permanent and reliable basis. All these excitements were the natural results of the daring and enterprising character of the people. But now, having worked off their superabundant steam, they would be prepared to go ahead systematically, and develop those resources which they had hitherto neglected. It was a course of medical effervescence highly beneficial to the body politic. All morbid appetite for sudden wealth was now gone forever.

But softly, good friends! What rumor is this? Whence come these silvery strains that are wafted to our ears from the passes of the Sierra Nevada? What dulcet Æolian harmonies—what divine, enchanting ravishment is it

> "That with these raptures moves the vocal air?"

As I live, it is a cry of Silver! Silver in WASHOE! Not gold now, you silly men of Gold Bluff; you Kern Riverites; you daring explorers of British Columbia! But SILVER — solid, pure SILVER! Beds of it ten thousand feet deep! Acres of it! miles of it! hundreds of millions of dollars poking their backs up out of the earth ready to be pocketed!

Do you speak of the mines of Potosi or Golconda?" Do you dare to quote the learned Baron Von Tschudi on South America and Mexico? Do you refer me to the ransom of Atahualpa, the unfortunate Inca, in the days of Pizarro? Nothing at all, I assure you, to the silver mines of Washoe! "Sir," said my informant to me, in strict confidence, no later than this morning, "you may rely upon it, for I am personally acquainted with a brother of the gentleman whose most intimate friend saw the man whose partner has just come over the mountains, and he says there never was the like on the face of the earth! The ledges are ten thousand feet deep — solid masses of silver. Let us be off! Now is the time! A pack-mule, pick and shovel, hammer and frying-pan will do. You need nothing more. HURRAH FOR WASHOE!"

Kind and sympathizing reader, imagine a man who for six years had faithfully served his government and his country; who had never, if he knew himself intimately, embezzled a dollar of the public funds; who had resisted the seductive influences of Gold Bluff, Kern, and Frazer Rivers from the purest motives of patriotism; who scorned to abandon his post in search of filthy lucre —imagine such a personage cut short in his official career, and suddenly bereft of his per diem by a formal and sarcastic note of three lines from head-quarters; then fancy you hear him jingle the last of his federal emoluments in his pocket, and sigh at the ingratitude of republics. Would you not consider him open to any proposition short of murder or highway robbery? Would you be surprised if he accepted an invitation from Mr. Wise, the aeronaut, to take a voyage in a balloon? or the berth of assistant manager in a diving-bell? or joined the first expedition in search of the treasure buried by the Spanish galleon on her voyage to Acapulco in 1578? Then consider his position, as he stands musing upon the mutability of human affairs, when those strange and inspiring cries of Washoe fall upon his ears for the first time, with a realizing sense of their import. Borne on the wings of the wind from the Sierra Nevada; wafted through every street, lane, and alley of San Francisco; whirling around the drinking saloons, eddying over the counters of the banking offices, scattering up the dust among the Front Street merchants,

HURRAH FOR WASHOE!

O 2

arousing the slumbering inmates of the Custom-house—what man of enterprise could resist it? Washoe! The Comstock lead! The Ophir! The Central—The Billy Choller Companies, and a thousand others, indicating in trumpet-tones the high road to fortune! From the crack of day to the shades of night nothing is heard but Washoe. The steady men of San Francisco are aroused, the men of Front Street, the gunny-bag men, the brokers, the gamblers, the butchers, the bakers, the whisky-dealers, the lawyers, and all. The exception was to find a sane man in the entire city.

No wonder the abstracted personage already referred to was aroused from his gloomy reflections. A friend appealed to him to go to Washoe. The friend was interested there, but could not go himself. It was a matter of incalculable importance. Millions were involved in it. He (the friend) would pay expenses. The business would not occupy a week, and would not interfere with any other business.

CHAPTER II.

START FOR WASHOE.

NEXT day an advertisement appeared in the city papers respectfully inviting the public to commit their claims and investments to the hands of their fellow-citizen, Mr. Yusef Badra, whose long experience in government affairs eminently qualified him to undertake the task of geological research. He was especially prepared to determine the exact amount of silver contained in fossils. It would afford him pleasure to be of service to his friends and fellow-citizens. The public would be so kind as to address Mr. Badra, at Carson City, Territory of Utah.

This looked like business on an extensive scale. It read like business of a scientific character. It was a card drawn up with skill, and calculated to attract attention. I am proud to acknowledge that I am the author, and, furthermore (if you will consider the information confidential), that I am the identical agent referred to.

THE AGENCY.

Many good friends shook their heads when I announced my intention of visiting Washoe, and, although they designed going themselves as soon as the snow was melted from the mountains, they could not understand how a person who had so long retained his faculties unimpaired could give up a lucrative government office and engage in such a wild-goose chase as that. Little did they know of the brief but irritating document which I carried in my pocket, and for which I am determined some day or other to write a satire against our system of government. I bade them a kindly farewell, and on a fine evening, toward the latter part of March, took my departure for Sacramento, there to take the stage for Placerville, and from that point as fortune might direct.

My stock in trade consisted of two pair of blankets, a spare shirt, a plug of tobacco, a note-book, and a paint-box. On my arrival in Placerville I found the whole town in commotion. There was not an animal to be had at any of the stables without applying three days in advance. The stage for Strawberry had made its last trip in consequence of the bad condition of the road. Every hotel and restaurant was full to overflowing. The streets were blocked up with crowds of adventurers all bound for Washoe. The gambling and drinking saloons were crammed to suffocation with customers practicing for Washoe. The clothing stores were covered with placards offering to sell goods at ruinous sacrifices to Washoe miners. The forwarding houses and express offices were overflowing with goods and packages marked for Washoe. The grocery stores were making up boxes, bags, and bundles of groceries for the Washoe trade. The stables were constantly starting off passenger and pack trains for Washoe. Mexican *vaqueros* were driving headstrong mules through the streets on the road to Washoe. The newspapers were full of Washoe. In short, there was nothing but Washoe to be seen, heard, or thought of. Every arrival from the mountains confirmed the glad tidings that enormous quantities of silver were being discovered daily in Washoe. Any man who wanted a fortune needed only to go over there and pick it up. There was Jack Smith, who made ten thousand dollars the other day at a single trade; and Tom Jenkins, twenty thousand by right of discovery; and Bill Brown, forty thousand in the tavern business, and so on. Every body was getting rich "hand over fist." It was the place for fortunes. No man could go amiss.

I was in search of just such a place. It suited me to find a fortune ready made. Like Professor Agassiz, I could not afford to make money, but it would be no inconvenience to draw a check on the great Washoe depository for fifty thousand dollars or so, and proceed on my travels. I would visit Japan, ascend the Amoor River, traverse Tartary, spend a few weeks in Siberia, rest a day or so at St. Petersburg, cross through Russia to the Black Sea, visit Persia, Nineveh, and Bagdad, and wind up somewhere in Italy. I even began to look about the bar-rooms for a map in order to lay out the route more definitely, but the only map to be seen was De Groot's outline of the route from Placerville to Washoe. I went to bed rather tired after the excitement of the day, and somewhat surfeited with Washoe. Presently I heard a tap at the door; a head was popped through the opening:

"I say, Cap!"

"Well, what do you say?"

"Are you the man that can't get a animal for Washoe?"

"Yes; have you got one to sell or hire?"

"No, I hain't got one myself, but me and my pardner is going to walk there, and if you like you can jine our party."

"Thank you; I have a friend who is going with me, but I shall be very glad to have more company."

"All right, Cap; good-night."

The door was closed, but presently opened again:

"I say, Cap!"

"What now?"

"Do you believe in Washoe?"

"Of course; why not?"

"Well, I suppose it's all right. Good-night; I'm in." And my new friend left me to my slumbers.

But who could slumber in such a bedlam, where scores and hundreds of crack-brained people kept rushing up and down the passage all night, in and out of every room, banging the doors after them, calling for boots, carpet sacks, cards, cocktails, and toddies; while

"I SAY, CAP!"

amid the ceaseless din arose ever and anon that potent cry of "Washoe!" which had unsettled every brain. I turned over and over for the fiftieth time, and at length fell into an uneasy doze. A mountain seemed to rise before me. Millions of rats with human faces were climbing up its sides, some burrowing into holes, some rolling down into bottomless pits, but all labeled Washoe.

Soon the mountain began to shake its sides with suppressed laughter, and out of a volcano on the top burst sheets of flame, through which jumped ten thousand grotesque figures in the shape of dollars with spider legs, shricking with all their might, "Washoe! ho! ho! Washoe! ho! ho!"

Surely the sounds were wonderfully real. Tap, tap, at the door.

"I say, Cap!"

"Well, what is it?"

"'Bout time to get up, if you calklate to make Pete's ranch to-night."

So I got up, and, after a cup of coffee, took a ramble on the heights, where I was amply compensated for my loss of rest by the richness and beauty of the sunrise. It was still early spring; the hills were covered with verdure; flowers bloomed in all directions; pleasant little cottages, scattered here and there, gave a civilized aspect to the scene; and when I looked over the busy town, and heard the lively rattle of stages, wagons, and buggies, and saw the long pack-trains winding their way up the mountains, I felt proud of California and her people. There is not a prettier little town in the state than Placerville, and certainly not a better class of people any where than her thriving inhabitants. They seemed, indeed, to be so well satisfied with their own mining prospects that they were the least excited of the crowd on the subject of the new discoveries. The impulse given to business in the town, however, was well calculated to afford them satisfaction. This was the last dépôt of trade on the way to Washoe. My excellent friend Dan Gelwicks, of the *Mountain Democrat*, assured me that he was perfectly satisfied to spend the remainder of his days in Placerville. Who that has ever visited the mountains, or attended a political convention in Sacramento, does not know the immortal "Dan"—the truest, best-hearted, handsomest fellow in existence; the very cream and essence of a country editor; who dresses as he pleases, chews tobacco when he pleases, writes tremendous political philippics, knows every body, trusts every body, sets up his own editorials, and on occasions stands ready to do the job and press work! I am indebted to "Dan" for the free use of his sanctum; and in consideration of his kindness and hospitality, do hereby transfer to him all my right, title, and interest in the Roaring Jack Claim, Wild-Cat Ledge, Devil's Gate, which by this time must be worth ten thousand dollars a foot.

Before we were quite ready to start our party had increased to five; but as each had to purchase a knife, tin cup, pound of cheese, or some other article of luxury, it was ten o'clock before we got fairly under way. And here I must say that, although our appearance as we passed along the main street of Placerville elicited no higher token of admiration than "Go it, Washoe!" such a party, habited and accoutred as we were, would have made a profound sensation in Hyde Park, London, or even on Broadway, New York.

The road was in good condition, barring a little mud in the neighborhood of "Hangtown;" and the day was exceedingly bright and pleasant. As I ascended the first considerable elevation in the succession of heights which extend all the way for a distance of fifty miles to the summit of the Sierra Nevada, and cast a look back over the foot-hills, a more glorious scene of gigantic forests, open valleys, and winding streams seldom greeted my vision. The air was singularly pure and bracing; every draught of it was equal to a glass of sparkling

"GO IT, WASHOE."

Champagne. At intervals, varying from fifty yards to half a mile, streams of water of crystal clearness and icy coolness burst from the mountain sides, making a pleasant music as they crossed the road. Whether the day was uncommonly warm, or the exercise rather heating, or the packs very heavy, it was beyond doubt some of the party were afflicted with a chronic thirst, for they stopped to drink at every spring and rivulet on the way, giving rise to a suspicion in my mind that they had not been much accustomed to that wholesome beverage of late. This suspicion was strengthened by a mysterious circumstance. I had lagged behind at a turn of the road to adjust my pack, when I was approached by the unique personage whose head in the doorway had startled me the night before.

"I say, Cap!" At the same time pulling from the folds of his blanket a dangerous-looking "pocket pistol," he put the muzzle to his mouth, and discharged the main portion of the contents down his throat.

"What d'ye say, Cap?"

Now I claim to be under no legal obligation to state what I said or did on that occasion; but this much I am willing to avow, that upon resuming our journey there was a glorious sense of freedom and independence in our adventurous mode of life. The fresh air, odorous with the scent of pine forests and wild flowers; the craggy rocks overhung with the grape and the morning-glory; the merry shouts of the Mexican *vaqueros*, mingled with the wild dashing of the river down the cañon on our right; the free exercise of every muscle; the consciousness of exemption from all farther restraints of office, were absolutely inspiring. I think a lyrical poem would not have exceeded my powers on that occasion. Every faculty seemed invigorated to the highest pitch of perfection. Hang the dignity of office! A murrain upon party politicians and inspector generals! To the bottomless pit with all vouchers, abstracts, and accounts current! I scorn that meagre and brainless style of the

THE POCKET PISTOL.

heads of the Executive Departments, "SIR,—Your services are no longer—" What dunce could not write a more copious letter than that? Who would be a slave when all nature calls upon him in trumpet tones to be free? Who would sell his birthright for a mess of pottage when he could lead the life of an honest miner —earn his bread by the sweat of his brow—breathe the fresh air of heaven without stint or limit? And of all miners in the world, who would not be a Washoe miner? Beyond question, this was a condition of mind to be envied and admired; and, notwithstanding the two pair of heavy blankets on my back, and a stiff pair of boots on my feet that galled my ankles most grievously, I really felt lighter and brighter than for years past. Nor did it seem surprising to me then that so many restless men should abandon the haunts of civilization, and seek variety and freedom in the wilderness of rugged mountains comprising the mining districts of the Sierra Nevada. The life of the miner is one of labor, peril, and exposure; but it possesses the fascinating element of liberty, and the promise of unlimited reward. In the midst of privations, amounting, at times, to the verge of starvation, what glowing visions fill the mind of the toiling adventurer! Richer in anticipation than the richest of his fellow-beings, he builds golden palaces, and scatters them over the world with a princely hand. He may not be a man of imagination; but in the secret depths of his soul there is a latent hope that some day or other he will strike a "lead," and who knows but it may be a solid mountain of gold, spangled with diamonds?

The road from Placerville to Strawberry Flat is for the most part graded, and no doubt is a very good road in summer; but it would be a violation of conscience to recommend it in the month of April. The melting of the accumulated snows of the past winter had partially washed it away, and what remained was deeply furrowed by the innumerable streams that sought an outlet in the ravines. In many places it seemed absolutely impracticable for wheeled vehicles; but it is an article of faith with California teamsters that wherever a horse can go a wagon can follow. There were some exceptions to this rule, however, for the road was literally lined with broken-down stages, wagons, and carts, presenting every variety of aspect, from the general smash-up to the ordi-

CALIFORNIA STAGE-DRIVER.

nary capsize. Wheels had taken rectangular cuts to the bottom; broken tongues projected from the mud; loads of dry-goods and whisky-barrels lay wallowing in the general wreck of matter; stout beams cut from the road-side were scattered here and there, having served in vain efforts to extricate the wagons from the oozy mire. Occasionally these patches of bad road extended for miles, and here the scenes were stirring in the highest degree. Whole trains of pack-mules struggled frantically to make the transit from one dry point to another; "burros," heavily laden, were frequently buried up to the neck, and had to be hauled out by main force. Now and then an enterprising mule would emerge from the mud, and, by attempting to keep the edge of the road, lose his foothold, and go rolling to the bottom of the cañon, pack and all. Amid the confusion worse confounded, the cries and maledictions of the *vaqueros* were perfectly overwhelming; but when the mules stuck fast in the mud, and it became necessary to unpack them, then it was that the *vaqueros* shone out most luminously. They shouted, swore, beat the mules, kicked them, pulled them, pushed them, swore again; and when all these resources failed, tore their hair, and resorted to prayer and meditation. Opposite is a faint attempt at the *vaquero* sliding-scale.

It will doubtless be a consolation to some of these unhappy *vaqueros* to know that such of their mules as they failed to extricate from the mud during the winter may, during the approaching summer, find their way out through the cracks. Should any future traveler be overtaken by thirst, and see a pair of ears growing out of the road, he will be safe in digging there, for underneath stands a mule, and on the back of that mule is a barrel of whisky.

WHISKY BELOW.

CARAMBO!—CARAJA—SACRAMENTO!—SANTA MARIA!—DIAVOLO!

Owing to repeated stoppages on the way, night overtook us at a place called "Dirty Mike's." Here we found a ruinously dilapidated frame shanty, the bar, of course, being the main feature. Next to the bar was the public bedroom, in which there was every accommodation except beds, bedding, chairs, tables, and washstands; that is to say, there was a piece of looking-glass nailed against the window-frame, and the general comb and tooth-brush hanging by strings from a neighboring post.

A very good supper of pork and beans, fried potatoes, and coffee, was served up for us on very dirty plates, by Mike's cook; and after doing it ample justice, we turned in on our blankets and slept soundly till morning. It was much in favor of our landlord that he charged us only double the customary price. I would cheerfully give him a recommendation if he would only wash his face and his plates once or twice a week.

The ascent of the mountains is gradual and continuous the entire distance to Strawberry. After the first day's journey there is but little variety in the scenery. On the right, a fork of the American River plunges down through a winding cañon, its force and volume augmented at short intervals by numerous smaller streams that cross the road, and by others from the opposite side. Thick forests of pine loom up on each side, their tops obscuring the sky. A few patches of snow lay along our route on the first day, but on the second snow was visible on both sides of the cañon.

The succession of scenes along the road afforded us constant entertainment. In every gulch and ravine a tavern was in process of erection. Scarcely a foot of ground upon which man or beast could find a foothold was exempt from a claim. There were even bars with liquors, offering a tempting place of refreshment to the weary traveler where no vestige of a house was yet perceptible. Board and lodging signs over tents not more than ten feet square were as common as blackberries in

BOARD AND LODGING.

June; and on no part of the road was there the least chance of suffering from the want of whisky, dry-goods, or cigars.

An almost continuous string of Washoeites stretched "like a great snake dragging its slow length along" as far as the eye could reach. In the course of this day's tramp we passed parties of every description and color: Irishmen, wheeling their blankets, provisions, and mining implements on wheel-barrows; American, French, and German foot-passengers, leading heavily-laden horses, or carrying their packs on their backs, and their picks and shovels slung across their shoulders; Mexicans, driving long trains of pack-mules, and swearing fearfully, as usual, to keep them in order; dapper-looking gentlemen, apparently from San Francisco, mounted on fancy horses; women, in men's clothes, mounted on mules or "burros;" Pike County specimens, seated on piles of furniture and goods in great lumbering wagons; whisky-peddlers, with their bar-fixtures and whisky on muleback, stopping now and then to quench the thirst of the toiling multitude; organ-grinders, carrying their organs; drovers, riding, raving, and tearing away frantically through the brush after droves of self-willed cattle designed for the shambles; in short, every imaginable class, and every possible species of industry, was represented in this moving pageant. It was a striking and impressive spectacle to see, in full competition with youth and strength, the most pitiable specimens of age and decay—white-haired old men, gasping for breath as they dragged their palsied limbs after them in the exciting race of avarice; cripples and hunchbacks; even sick men from their beds—all stark mad for silver.

But the tide was not setting entirely in the direction of Carson Valley. A counter-current opposed our progress in the shape of saddle-trains without riders, long lines of pack-mules laden with silver ore, scattering parties of weather-beaten and foot-sore pedestrians, bearing their hard experience in their faces, and solitary stragglers, of all ages and degrees, mounted on skeleton horses, or toiling wearily homeward on foot—some merry, some sad, some eagerly intent on farther speculation, but all bearing the unmistakable impress of Washoe.

Among the latter, a lank, leathery-looking fellow, doubtless from the land of wooden nutmegs, was shambling along through the mud, talking to himself appar ently for want of more congenial fellowship. I was about to pass him, when he arrested my attention:

"Look here, stranger!"

I looked.

"You're bound for Washoo, I reckon?"

I was bound for Washoe.

"What line of business be you goin' into there?"

Was not quite certain, but thought it would be the agency line.

GRINDSTONES.

"Ho! the agency line—stage-agent, maybe? Burche's line, I guess?"

That was not it exactly; but no matter. Perhaps I could do something for him in Washoe.

"Nothing, stranger, except to keep dark. Do you know the price of grindstones in Placerville?"

I didn't know the price of grindstones in Placerville, but supposed they might be cheap, as there were plenty there.

"That's my hand exactly!" said my friend, with an inward chuckle of satisfaction. I expressed some curiosity to know in what respect the matter of grindstones suited his hand so well, when, looking cautiously around, he drew near, and informed me confidentially that he had struck a "good thing" in Washoe. He had only been there a month, and had made a considerable pile. There was a dreadful scarcity of grindstones there, and, seeing that miners, carpenters, and mechanics of all sorts were hard up for something to sharpen their tools on, he had secured the only grindstone that could be had, which was pretty well used up when he got it. But he rigged it up ship-shape and Bristol fashion, and set up a grinding business, which brought him in from twenty to thirty dollars a day, till nothing was left of the stone. Now he was bound to Placerville in search of a good one, with which he intended to return immediately. I wished him luck and proceeded on my way, wondering what would turn up next.

It was not long before I was stopped by another enterprising personage; but this was altogether a different style of man. There was something brisk and spruce in his appearance, in spite of a shirt far gone in rags and a shock of hair that had long been a stranger to the scissors. What region of country he came from it was impossible to say. I think he was a cosmopolite, and belonged to the world generally.

"Say, Colonel!"—this was his style of address—"on the way to Washoe?"

A SPECULATOR.

"Yes."

"Excuse me: I have a little list of claims here, Colonel, which I would like to show you;" and he pulled from his shirt-pocket a greasy package of papers, which he dexterously unfolded. "Guess you're from San Francisco, Colonel? Here is—let me see—

200 feet in the Pine Nut,
300 feet in the Grizzly Ledge,

150 feet in the Gouge Eye,
125 feet in the Wild-Cat,
100 feet in the Root-Hog-or-Die,
50 feet in the Bobtail Horse,
25 feet in the Hell Roaring;

and many others, Colonel, in the best leads. Now the fact is, d'ye see, I'm a little hard up, and want to make a raise. I'll sell all, or a part, at a considerable sacrifice for a small amount of ready cash."

"How much do you want?"

"Why, if I could raise twenty dollars or so, it would answer my present purpose; I'll sell you twenty feet in any of these claims for that amount. Every foot of them is worth a thousand dollars; but, d'ye see, they're not yet developed."

Circumstances forced me to decline this offer, much to the disgust of the enterprising speculator in claims, who assured me I might go farther and fare worse; but somehow the names did not strike me as attractive in a mineral point of view.

I had by this time lost the run of all my comrades, and was obliged to pursue my journey alone. Three had gone ahead, and the other was nearly used up. The day had opened fairly, but now there were indications of bad weather. It was quite dark when I reached a small shanty about four miles from Strawberry. Here I halted till my remaining comrade came up. The proprietor of the shanty was going into the tavern business, and was engaged in building a large clapboard house. His men were all at supper, and in reply to our application for lodgings, he told us we might sleep in the calf-pen if we liked, but there was no room in the house. He could give us something to eat after his workmen were done supper, but not before. He had brandy and gin, but no tea to spare. On the whole, he thought we had better go on to Strawberry.

Now this was encouraging. It was already pattering down rain, and the calf-pen to which he directed us was knee-deep in mud and manure, without roof or shelter of any kind. Even the unfortunate progeny of the old cow, which ran bellowing around the fence, in motherly solicitude for her offspring, shivered with cold, and made piteous appeals to this hard-hearted man. I finally bribed him, by means of a gold dollar, to let us have a small piece of bread and a few swallows of tea. Thus refreshed, we resumed our journey.

Four miles more of slush and snow, up hill nearly all the way, across rickety bridges, over roaring cataracts, slippery rocks, stumps, and brush, through acres of black oozy mire, and so dark a bat could scarcely recognize his own father! It was a walk to be remembered. The man in the shanty, if he possess a spark of humanity, will, I trust, feel bitterly mortified when he reads this article. He caused me some gloomy reflections upon human nature, which have been a constant source of repentance ever since. But consider the provocation. The rain poured down heavily, mingled with a cutting sleet; a doleful wind came moaning through the pines; our blankets were wet through, and not a stitch upon our backs left dry; even my spare shirt was soaking the strength out of the plug of tobacco so carefully stowed away in its folds, and my paints were giving it what aid they could in the way of color.

Well, there is an end to all misery upon earth, and so there was to this day's walk. A light at length glimmered through the pines, first faint and flickering, then a full blaze, then half a dozen brilliant lights, which proved to be camp-fires under the trees, and soon we stood in front of a large and substantial log house. This was the famous "Strawberry," known throughout the length and breadth of the land as the best stopping-place on the route to Washoe, and the last station before crossing the summit of the Sierra Nevada. The winter road for wheel-vehicles here ended; and, indeed, it may be said to have ended some distance below, for the last twelve miles of the road seemed utterly impracticable for wagons. At least, most of those I saw were fast in the mud, and likely to remain there till the beginning of summer. Dark and rainy as it was, there were crowds scattered around the house, as if they had some secret and positive enjoyment in the contemplation of the weather. Edging our way through, we found the bar-room packed as closely as it could be without bursting out some of the walls; and of all the motley gangs that ever happened together within a space of twenty feet, this certainly was the most extraordinary and the most motley. Dilapidated gentlemen with slouched hats and big boots, Jew peddlers dripping wet, red-shirted miners, teamsters, vaqueros, packers, and traders, swearing horribly at nothing; some drinking at the bar, some warming themselves before a tremendous log fire that sent up a reeking steam from the conglomerated mass of wet and muddy clothes, to say nothing of the boots and socks that lay simmering near the coals. A few bare and sore footed outcasts crouched down in the corners, trying to catch a nap, and here and there a returned Washoeite, describing in graphic language, garnished with oaths, the wonders and beauties of Virginia City. But chiefly remarkable in the crowd was the regiment of light infantry, pressed in double file against the dining-room door, awaiting the fourth or fifth charge at the table.

At the first tinkle of the bell the door was burst open with a tremendous crash, and for a moment no battle-scene in Waterloo, no charge at Resaca de la Palma or the heights of Chapultepec, no Crimean avalanche of troops dealing death and destruction around them, could have equaled the terrific onslaught of the gallant troops of Strawberry. The whole house actually tottered and trembled at the concussion, as if shaken by an earthquake. Long before the main body had assaulted the table the din of arms was heard above the general uproar; the deafening clatter of plates, knives, and forks, and the dreadful battle-cry of "Waiter! waiter! Pork

DINNER AT STRAWBERRY.

P 2

and beans! Coffee, waiter! Beefsteak! Sausages! Potatoes! Ham and eggs—quick, waiter, for God's sake!" It was a scene of destruction and carnage long to be remembered. I had never before witnessed a battle, but I now understood how men could become maddened by the smell of blood. When the table was vacated it presented a shocking scene of desolation. Whole dishes were swept of their contents; coffee-pots were discharged to the dregs; knives, forks, plates, and spoons lay in a confused mass among the bones and mutilated remnants of the dead; chunks of bread and hot biscuit were scattered broadcast, and mince-pies were gored into fragments; tea-cups and saucers were capsized; and the waiters, hot, red, and steamy, were panting and swearing after their superhuman labors.

Half an hour more and the battle-field was again cleared for action. This was the sixth assault committed during the evening; but it was none the less terrible on that account. Inspired by hunger, I joined the army of invaders this time, and by gigantic efforts of strength maintained an honorable position in the ranks. As the bell sounded, we broke! I fixed my eye on a chair, rushed through the struggling mass, threw out my hands frantically to seize it, but, alas! it was already captured. A dark-visaged man, who looked as if he carried concealed weapons on his person, was seated in it, shouting hoarsely the battle-cry of "Pork and beans! Waiter! Coffee, waiter!" Up and down the table it was one gulping mass, jaws distended, arms stretched out, knives, forks, and even the bare hands plunged into the enemy. Not a spot was vacant. I venture to assert that from the commencement of the assault till the capture and complete investment of the fortifications did not exceed five seconds. The storming of the Malakoff and the fall of Sebastopol could no longer claim a place in history.

At length fortune favored the brave. I got a seat at the next onslaught, and took ample satisfaction for the delay by devouring such a meal as none but a hardy Washoeite could be expected to digest. Pork and beans, cabbage, beef-steak, sausages, pies, tarts, coffee and tea, eggs, etc.—these were only a few of the luxuries furnished by the enterprising proprietor of the "Strawberry." May every blessing attend that great benefactor of mankind! I say it in all sincerity; he is a great and good man, a Websterian innkeeper, for he thoroughly understands the constitution. I would give honorable mention to his name if I knew it; but it matters not; his house so far surpasses the Metropolitan or the St. Nicholas that there is no comparison in the relish with which the food is devoured. In respect to sleeping accommodations there may be some difference in their favor. I was too late to secure a bed in the general bedroom up stairs, where two hundred and fifty tired wayfarers were already snoring in double-shotted bunks 2×6; but the landlord was a man of inexhaustible resources. A private whisper in his ear made him a friend forever. He nodded sagaciously, and led me into a small parlor about 15×20, in which he gave my company of five what he called a "lay-out," that is to say, a lay-out on the floor, with our own blankets for beds and covering. This was a special favor, and I would have cherished it in my memory for years had not a suspicion been aroused in my mind before the lapse of half an hour that there were others in the confidence of mine host. Scarcely had I entered upon the first nap when somebody undertook to walk upon me, commencing on my head and ending on the pit of my stomach. I grasped him firmly by the leg. He apologized at once in the most abject manner; and well for him he did, for it was enough to incense any man to be suddenly roused up in that manner. The intruder, I discovered, was a Jew peddler. He offered me a cigar, which I smoked in token of amity; and in the mean time he turned in alongside and smoked another. When daylight broke I cast around me to see what every body was doing to create

THE "LAY OUT."

such a general commotion. I perceived that there were about forty sleepers, all getting up. Boots strongly scented with feet, and stockings of every possible degree of odor, were lying loose in all directions; blankets, packs, old clothes, and ragged shirts, and I don't know what all—a palpable violation of the landlord's implied compact. True, he had not agreed to furnish a single bed for five, but he never hinted that he was going to put forty men, of all sorts and sizes, in the same general "lay-out," as he was pleased to style it, and that only

THE STOCKING-THIEF.

large enough for half the number. Once, in Minnesota, I slept in a bed with eight, and gave considerable offense to my landlord when I remonstrated against his putting in a ninth. He said he liked to see a man "accommodating"—a reflection upon my good-nature which I considered wholly unwarranted by the circumstances. But this was even a stronger case.

The Jew peddler had not undressed, and, not to judge him harshly, I don't think he ever did undress. He was soon up, and left, as I suppose, while I was dressing. With him departed my stockings. They were not very fine—perhaps, considering the muddy road, not very clean; but they were all I had, and were valuable beyond gold or silver in this foot-weary land. I never saw them more. What aggravated the offense, when I came to review it seriously, was, that I remembered having seen him draw just such a pair over his boots, as a protection against the snow, without the remotest suspicion of the great wrong he was doing me.

We shall meet this Stocking-thief again.

CHAPTER III.

ACROSS THE MOUNTAINS.

UPON taking an observation from the front door at Strawberry, we were rather startled to find that the whole place was covered with snow to the depth of two or three feet. The pack trains had given up all hope of getting over the mountain. It was snowing hard, and the appearance of the weather was dark and threatening. To be housed up here with three or four hundred men, and the additional numbers that might be expected before night, was not a pleasant prospect; but to be caught in a snow-storm on the summit, where so many had perished during the past winter, was worse still. Upon reviewing the chances I resolved to start, and if the storm continued I thought there would be no difficulty in finding the way back. It was eight miles of a continuous and precipitous ascent to the summit, and three miles from that point to the Lake House in Lake Valley, where the accommodations were said to be the worst on the whole trail.

A few miles from Strawberry one of the party gave out in consequence of sore feet; the other two pushed

THE TRAIL FROM STRAWBERRY.

on, despite the storm which now raged fearfully, but had not proceeded far when they were forced to turn back. I was loth to leave my disabled friend, and returned with him to Strawberry, where we had a repetition of nearly all that has already been described, only a little intensified in consequence of increased numbers. The others of our party stopped somewhere on the road, and I did not meet them again until next afternoon at Woodford's, on the other side of the mountain.

As soon as it was light next morning I took another observation of the weather. It was still snowing, but not so heavily as on the preceding day. My remaining partner was by this time completely crippled in his feet, and had to hire a horse at the rate of twenty dollars for twenty-five miles.

I was delayed some hours in getting off, owing to the pressure of the forces at the breakfast-table, but finally made a fair start for the summit. My pack had become a source of considerable inconvenience. I was accustomed to walking, but not to carrying a burden of twenty or twenty-five pounds. My shoulders and ankles were so galled that every step had to be made on the nicest calculation; but the new snow on top of the old trail began to melt as soon as the sun came out, making a very bad trail for pedestrians. Two miles from Strawberry we crossed a bridge, and struck for the summit.

Here we had need of all our powers of endurance. It was a constant struggle through melted snow and mud—slipping, sliding, grasping, rolling, tumbling, and climbing, up again and still up, till it verily seemed as if we must be approaching the clouds. The most prominent peculiarity of these mountains is, that a person on foot, with a heavy load on his back, is never at the top when he imagines he is; the "divide" is always a little farther on and a little higher up—at least until he passes it, which he does entirely ignorant of the fact. There is really no perceptible "divide;" you pass a series of elevations, and commence the descent without any apparent difference in the trail.

The pack trains had broken through the old snow in many places, leaving deep holes, which, being now partially covered with recent snow, proved to be regular man-traps, often bringing up the unwary pedestrian "all standing." The sudden wrenching of the feet in the smaller holes, which had been explored by the legs of horses, mules, and cattle, was an occurrence of every ten or a dozen steps. In many places the trail was perfectly honeycombed with holes, where the heavily-laden animals had cut through the snow, and it was exceedingly difficult to find a foothold. To step on either side and avoid these bad places would seem easy enough, but I tried it on more than one occasion, and got very nearly buried alive. All along the route, at intervals of a mile or two, we continued to meet pack trains; and as every body had to give way before them, the tumbling out and plunging in the snow were very lively.

I walked on rapidly in the hope of making Woodford's—the station on the eastern slope of the mountain—before night, and by degrees got ahead of the main body of footmen who had left Strawberry that morning. In a narrow gorge, a short distance from the commencement of the descent into Lake Valley, I happened to look up a little to the right, where, to my astonishment, I perceived four large brown wolves sitting on their haunches not over twenty feet from me! They seemed entirely unconcerned at my presence, except in so far as they may have indulged in some speculation as to the amount of flesh contained on my body. As I was entirely unarmed, I thought it would be but common politeness to speak to them, so I gave them a yell in the Indian language. At this they retired a short distance, but presently came back again as if to inquire the exact meaning of my salutation. I now thought it best not to be too intimate, for I saw they were getting rather familiar on a short acquaintance; and picking up a stick of wood, I made a rush and a yell at them which must have been formidable in the extreme. This time they

"WE ARE WAITING FOR YOU."

retreated more rapidly, and seemed undecided about returning. At this crisis in affairs a pack train came along, the driver of which had a pistol. Upon pointing out the wolves to him he fired, but missed them. They then retreated up the side of the mountain, and I saw nothing more of them.

The descent of the "grade" was the next rough feature in our day's journey. From the point overlooking Lake Valley the view is exceedingly fine. Lake Bigler —a sheet of water forty or fifty miles in length by ten or fifteen wide—lies embosomed in the mountains in full view from this elevation; but there was a drizzling sleet which obscured it on this occasion. I had a fine sight of it on my return, however, and have seldom witnessed any scene in Europe or elsewhere to compare with it in extent and grandeur.

The trail on the grade was slippery with sleet, and walking upon it was out of the question. Running, jumping, and sliding were the only modes of locomotion at all practicable. I tried one of the short cuts, and found it an expeditious way of getting to the bottom.

A SHORT CUT.

Some trifling obstruction deprived me of the use of my feet at the very start, after which I traveled down in a series of gyrations at once picturesque and complicated. When I reached the bottom I was entirely unable to comprehend how it had all happened; but there I was, pack and baggage, all safely delivered in the snow—bones sound, and free of expense.

At the Lake House—a tolerably good-sized shanty at the foot of the grade—we found a large party assembled, taking their ease as they best could in such a place, without much to eat and but little to drink, except old-fashioned tarentula-juice, "warranted to kill at forty paces."

The host of the Lake was in a constant state of nervous excitement, and did more scolding, swearing, gouging, and general hotel work in the brief space of half an hour than any man I ever saw. He seemed to be quite worn out with his run of customers—from a hundred to three hundred of a night, and nowhere to stow 'em—all cussin' at him for not keepin' provisions; and how could he, when they ate him clean out every day, and some of 'em never paid him, and never will?

I was not sorry to get clear of the Lake House, its filth, and its troubles.

Upon crossing the valley, which is here about a mile wide, the ascent of the next summit commences. Here we had almost a repetition of the main summit, except that the descent on the other side is more gradual.

At length we struck the beginning of Hope Valley. I shall always remember this portion of the journey as the worst I ever traveled on foot. Every yard of the trail was honeycombed to the depth of two or three feet. On the edges there was no foothold at all; and occasionally we had to wade knee-deep in black, sticky mire, from which it was difficult to extricate one's feet and boots at the same time. I was glad enough when myself and two casual acquaintances succeeded in reaching the solitary log house which stands near the middle of the valley.

I little expected to find in this wilderness a philosopher of the old school; but here was a man who had evidently made up his mind to withstand all the allurements of wealth, and devote the remainder of his life to ascetic reflections upon the follies of mankind. Diogenes in his tub was not more rigorous in his seclusion than this isolated inhabitant of Hope Valley. His log cabin, to be sure, was some improvement, in extent, upon the domicile of that famous philosopher; but in point of architectural style, I don't know that there could have been much advantage either way.

A few empty bags, and a bar entirely destitute of bottles, with a rough bench to sit upon, comprised all the furniture that was visible to the naked eye. From a beam overhead hung a bunch of foxskins, which emitted a very gamy odor; and the clay floor had apparently never been swept, save by the storms that had passed over it before the cabin was built. A couple of rifles hung upon pegs projecting from the chimney, and a powder-flask was the only mantle-piece ornament. Diogenes sat, or rather reclined, on the pile of empty sacks, holding by the neck a fierce bull-dog. The sanguinary propensities of this animal were manifested by repeated attempts to break away, and seize somebody by the throat or the leg; not that he growled, or snarled, or showed any puppyish symptoms of a trifling kind, but there was a playful switching of his tail and a leer of the eye uncommonly vicious and tiger-like. It certainly would not have taken him more than two minutes to hamstring the stoutest man in the party.

Between the dog and his master there was a very striking congeniality of disposition, if one might judge by the expression of their respective countenances. It would apparently have taken but little provocation to make either of them bite.

Battered and bruised as we were, and hungry into the bargain, after our hard struggle over the mountain, it became a matter of vital importance that we should secure lodgings for the night, and, if possible, get something to eat. The place looked rather unpromising; but, after our experience in Lake Valley, we were not easily discouraged. Upon broaching the subject to Diogenes in the mildest possible manner, his brow darkened, as if a positive insult to his common sense had been attempted.

DIOGENES.

"Stay here all night!" he repeated, savagely. "What the h—ll do you want to stay here all night for?"

We hinted at a disposition to sleep, and thought he might possibly have room on the floor for our blankets.

At this he snapped his fingers contemptuously, and muttered, "Can't come that over me! I've been here too long for that!"

"But we are willing to pay you whatever is fair."

"Pay? Who said I wanted pay? Do I look like a man that wants money?"

We thought not.

"If I wanted money," continued Diogenes, "I could have made fifty dollars a day for the last two months. But I ask no favors of the world. Some of 'em wants to stay here whether I will or no; I rather think I'm too many for any of that sort—eh, Bull, what d'ye say?" Bull growled, with a bloodthirsty meaning. "Too many altogether, gents—me and Bull."

There was a sturdy independence about this fellow, and a scorn for filthy lucre that rather astonished me as a citizen of a money-loving state.

"Well, if you can't let us stay all night, perhaps you can get us up a snack of dinner?"

"Snack of dinner?"—and here there was a guttural chuckle that boded failure again—"I tell you this ain't a tavern; and if it was, my cook's gone out to take a airing."

"But have you nothing in the house to eat?"

"Oh yes, there's a bunch of foxskins. If you'd like some of 'em cooked, I'll bile 'em for you."

This man's disposition had evidently been soured in early life. I think he must have been crossed in love. His style had the merit of being terse, but his manner was sarcastic to the verge of impoliteness.

"Well, I suppose we can warm ourselves at the fire?"

"If you can," quoth Diogenes, "you can do more than I can; and here he hauled his blanket over his shoulders, and fell back on the empty potato sacks as if there was no more to be said on that or any other subject.

The bull-dog seemed to be of the same way of thinking, and quietly laid down by his master; still, however, keeping his eye on us, as suspicious characters.

Nothing remained but to push on for Woodford's, distant six miles.

Now, when you come to put six miles on the end of

a day's journey such as ours had been, it becomes a serious matter. Besides, it was growing late, and a terrific wind, accompanied by a blinding sleet, rendered it scarcely practicable to stand up, much less to walk. I do not know how we ever staggered over that six miles. The last three, however, were down hill, and not so bad, as the snow was pretty well gone from the cañon on the approach to Woodford's.

This is the last station on the way over from Carson, and forms the upper terminus of that valley. It is supposed to be in Utah, but our landlord could not tell us exactly where the boundary-line ran.

We found here several hundred people, bound in both directions, and passed a very rough night, trying to get a little sleep amid the motley and noisy crowd.

I had endured the journey thus far very well, and had gained considerably in strength and appetite. The next day, however, upon striking into the sand of Carson Valley, my feet became terribly blistered, and the walking was exceedingly painful. There are some good farms in the upper part of the valley, between Woodford's and Genoa, though the general aspect of the country is barren in the extreme.

By sundown I had made only fifteen miles, and still was three miles from Genoa. Every hundred yards was now equal to a mile. At length I found it utterly impossible to move another step. It was quite dark, and there was nothing for it but to sit down on the road-side. Fortunately, the weather was comparatively mild. As I was meditating how to pass the night, I perceived a hot spring close by, toward which I crept; and finding the water strongly impregnated with salt, it occurred to me that it might benefit my feet. I soon plunged them in, and in half an hour found them so much improved that I was enabled to resume my journey. An hour more, and I was snugly housed at Genoa.

This was a place of some importance during the time of the Mormon settlements, but had not kept pace with

Carson City in the general improvement caused by the recent discoveries. At present it contained a population of not more than two or three hundred, chiefly store-keepers, teamsters, and workmen employed upon a neighboring saw-mill. The inhabitants professed to be rich in silver leads, but upon an examination of the records to find the lead in which my San Francisco friend had invested, and which was represented to be in this district, I was unable to find any trace of it; and there was no such name as that of the alleged owner known or ever heard of in Genoa. In fact, as I afterward ascertained, it was purely a fictitious name, and the whole transaction was one of those Peter Funk swindles so often practiced upon the unwary during this memorable era of swindles. I don't know how my friend received the intelligence, but I reported it to him without a solitary mitigating circumstance. Had I met with the vile miscreant who had imposed upon him, I should have felt bound to resort to personal measures of satisfaction, in consideration of the fund expended by my friend on the expenses of this commission of inquiry. The deeds were so admirably drawn, and the names written so legibly, that I don't wonder he was taken in. In fact, the only obstacle to his scheme of sudden wealth was, that there were no such mines, and no such men as the alleged discoverers in existence.

I proceeded the next day to Carson City, which I had fixed upon as the future head-quarters of my agency. The distance from Genoa is fifteen miles, the road winding around the base of the foot-hills most of the way. I was much impressed with the marked difference between the country on this side of the Sierra Nevada range and the California side. Here the mountains were but sparsely timbered; the soil was poor and sandy, producing little else than stunted sage bushes; and the few scattering farms had a thriftless and poverty-stricken look, as if the task of cultivation had proved entirely hopeless, and had long since been given up. Across the

CARSON CITY.

valley toward the Desert, ranges of mountains, almost destitute of trees, and of most stern and forbidding aspect, stretched as far as the eye could reach. Carson River, which courses through the plain, presented the only pleasing feature in the scene.

I was rather agreeably suprised at the civilized aspect of Carson City. It is really quite a pretty and thrifty little town. Situated within a mile of the foot-hills, within reach of the main timber region of the country, and well watered by streams from the mountains, it is rather imposing on first acquaintance; but the climate is abominable, and not to be endured. I know of none so bad except that of Virginia City, which is infinitely worse. The population was about twelve or fifteen hundred at the time of my visit. There was great speculation in town lots going on, a rumor having come from Salt Lake that the seat of government of Utah was about to be removed to Carson. Hotels and stores were in progress of erection all about the Plaza, but especially drinking and gambling saloons, it being an article of faith among the embryo sovereigns of Utah that no government can be judiciously administered without plenty of whisky, and superior accommodations for "bucking at monte." I am not sure but there is a similar feature in the California Constitution; at least, the practice is carried on to some extent at Sacramento during the sittings of the Legislature. Measures of the most vital importance are first introduced in rum cocktails, then steeped in whisky, after which they are engrossed in gin for a third reading. Before the final vote the opponents adjourn to a game of *poker* or *sledge*, and upon the amount of Champagne furnished on the occasion by the respective parties interested in the bill depends its passage or defeat. It was said that Champagne carried one of the great senatorial elections; but this has been denied, and it would be dangerous to insist upon it.

I had the pleasure of meeting in Carson an esteemed friend from San Francisco, Mr. A. J. Van Winkle, Real

Estate Agent, who, being a descendant of the famous Rip Van Winkle, was thoughtful enough to furnish me with a bunk to sleep in. Warned by the fate of his unhappy ancestor, my friend had gone briskly into the land business, and now owned enough of town lots, of amazingly appreciative value, to keep any man awake for the remainder of his life. I think if I had as much property, doubling itself up all the time like an acrobat in a circus, I would never sleep another wink thinking about it.

Chief among the curiosities of Carson City is the *Territorial Enterprise*—a newspaper of an origin long anterior to the mining excitement. I was introduced to "the Colonel," who presides over the editorial department, and found him uncommonly strong on the ultimate destiny of Carson. His office was located in a dirty frame shanty, where, amid types, rollers, composing-stones, and general rubbish of a dark and literary aspect, those astounding editorials which now and then arouse the public mind are concocted. The Colonel and his compositors live in a sort of family fashion, entirely free from the rigorous etiquette of such establishments in New York. They cook their own food in the composition room (which is also the editorial and press room), and being, as a general thing, short of plates, use the frying-pan in common for that purpose. In cases of great festivity and rejoicing, when a subscriber has settled up arrearages or the cash is paid down for a good job of hand-bills, the Colonel purchases the best tenderloin steak to be had in market, and cooks it with one hand, while with the other he writes a letter of thanks to the subscriber, or a puff on the hand-bill. But the great hope upon which the Colonel feeds his imagination is the removal of the seat of government from Salt Lake to Carson City, which he considers the proper place. Mr. Van Winkle is also of the same opinion; and, as a general thing, the proposition is favorably entertained by the citizens of Carson.

As usual in new countries, a strong feeling of rivalry

exists between the Carsonites and the inhabitants of Virginia City. I have summed up the arguments on both sides and reduced them to the following pungent essence:

Virginia City—a mud-hole; climate, hurricanes and snow; water, a dilution of arsenic, plumbago, and copperas; wood, none at all except sage-brush; no title to property, and no property worth having.

Carson City—a mere accident; occupation of the inhabitants, waylaying strangers bound for Virginia; business, selling whisky, and so dull at that, men fall asleep in the middle of the street going from one groggery to another; productions, grass and weeds on the Plaza.

While this fight is going on, Silver City, which lies about midway between the two, shrugs her shoulders and thanks her stars there can be no rivalry in her case. If ever there was a spot fitted by nature for a seat of government, it is Silver City—the most central, the most moral, the most promising; in short, the only place where the seat of government can exist for any length of time.

This Kilkenny-cat fight is highly edifying to a stranger, who, of course, is expected to take sides, or at once acknowledge himself an enemy. The result, I hope, will be satisfactory and triumphant to all parties. I would suggest that the government be split into three slices, and a slice stowed away under ground in each of the great cities, so that it may permeate the foundations of society.

CHAPTER IV.

AN INFERNAL CITY.

A FEW days after my arrival in Carson the sky darkened, and we soon had a specimen of the spring weather of this region. To say that it stormed, snowed, and rained would be ridiculously tame in comparison with the real state of the case. The wind whistled through the thin shanties in a manner that left scarcely a hope of roof or frame standing till night. Through the crevices came little hurricanes of snow-drift mixed with sand; each tenement groaned and creaked as if its last hour had come; the air was bitterly cold; and it seemed, in short, as if the vengeance of Heaven had been let loose on this desolate and benighted region.

Next day the clouds gradually lifted from the mountain tops, and the sun once more shone out bright and clear. The snow, which now covered the valley, began to disappear; the lowing of half-starved cattle, in search of the few green patches visible here and there, gave some promise of life; but soon the portentous gusts of wind swept down again from the cañons; dark clouds overspread the sky, and a still more violent storm than on the preceding day set in, and continued without intermission all night. By morning the whole face of the country was covered with snow. A few stragglers came in from Woodford's, who reported that the trail to Placerville was covered up to the depth of six or eight feet, and was entirely unpracticable for man or beast. Apprehensions were felt for the safety of the trains on the way through, as nothing could be heard from them. A large party had started out to open the trail, but were forced back by the severity of the weather. The snow-drifts were said to vary from twenty to thirty feet in depth.

Here was a pretty predicament! To be shut up in this desolate region, where even the cattle were dying of starvation, with seven or eight thousand human mouths to be fed, and the stock of provisions rapidly giving out, was rather a serious aspect of affairs. I do not know that actual starvation could have resulted for some time, certainly not until what cattle were alive had been killed, and soup made of the dead carcasses that covered the plain. Even before resorting to the latter extremity there were horses, mules, burros, and dogs on hand, upon which the cravings of hunger might be appeased for a month or so; and in the event of all these resources giving out, should the worst come to the worst, the few Digger Indians that hung around the settlements might be made available as an article of temporary subsistence.

In this extremity, when considerable suffering, if not absolute starvation, stared us in the face, the anxiety respecting the opening of the trails became general. Groups of men of divers occupations stood in the streets, or on every little rise of ground in the neighborhood, speculating upon the chances or peering through the gloom in the hope of discerning the approach of some relief train. The sugar was gone; flour was eighty dollars a sack, and but little to be had at that; barley was seventy-five cents a pound, and hay sixty cents; horses were dying for want of something to eat; cigars were rapidly giving out; whisky might stand the pull another week, but the prospect was gloomy of any thing more nourishing.

In this exciting state of affairs, when every brain was racked to devise ways and means of relief, and when hope of succor was almost at an end, a scout came running in from the direction of the Downerville trail with the glorious tidings of an approaching mule train. The taverns, billiard saloons, groggeries, and various stores were soon empty—every body rushed down the street to have assurance made doubly sure. Cheer after cheer burst from the elated crowd when the train hove in sight. On it came—at first like a row of ants creeping down the hillside; then nearer and larger, till the clatter of the hoofs and the rattling of the packs could be heard; then the blowing of the tired mules; and at last the leader, an old gray mule, came staggering wearily along heavily packed. A barrel was poised on his back—doubtless a barrel of beef, or it might be pork, or bacon. The brand heaves in sight. Per Baccho! it is neither beef, pork, nor bacon, but *whisky*—old Bourbon whisky! The next mule totters along under two half barrels. Speculation is rife. Every man with a stomach and an appetite for wholesome food is interested. Pigs' feet perhaps, or mackerel, or, it may be, preserved chicken? But here is the mark—*brandy*; by the powers! nothing but *brandy!* However, here comes the third with a load of five-gallon kegs—molasses beyond question, or lard, or butter? Wrong again, gentlemen—*gin*, nothing but *gin*. On staggers a fourth, heavily burdened with more kegs—sugar, or corn-meal, or preserved apples, I'll bet my head. Never bet your head. It is nothing but bitters—*Mack's Bitters!* But surely the fifth carries a box of crushed sugar on his back, he bears himself so gayly under his burden. And well he may! That box contains no more sugar than you do, my friend; it is stuffed choke-full with decanters, tumblers, and pewter spoons. But there are still ten or fifteen mules more. Surely there must be some provisions in the train. Nobody can live to a very protracted period of life on brandy, whisky, gin, Mack's Bitters, and glass-ware. Alas for human expectation! One by one the jaded animals pass, groaning and tottering under their heavy burdens—a barrel of rum; two boxes of bottled ale; six crates of Champagne; two pipes of California wine; a large crate of bar fixtures; and a dozen boxes of cigars—none of them nutritious articles of subsistence.

As if to enhance our troubles, the party in charge of the train had been nearly starved out in the mountains, and now came in the very lankest and hungriest of the crowd. If they were thirsty, it was their own fault; but none of them looked as if they had suffered in that respect.

Before entering into the responsible duties of my agency, I was desirous of seeing as much of the mining region as possible, and with this view took the stage for Virginia City. The most remarkable peculiarity on the road was the driver, whose likeness I struck in a happy moment of inspiration. At Silver City, eight miles from

THE STAGE.

Carson, I dismounted, and proceeded the rest of the way on foot. The road here becomes rough and hilly, and but little is to be seen of the city except a few tents and board shanties. Half a mile beyond is a remarkable gap cut by Nature through the mountain, as if for the express purpose of giving the road an opportunity to visit Virginia City.

As I passed through the Devil's Gate it struck me that there was something ominous in the name. "Let all who enter here—" But I had already reached the other side. It was too late now for repentance. I was about to inquire where the devil— Excuse me, I use the word in no indecorous sense. I was simply about to ask where he lived, when, looking up the road, I saw amid the smoke and din of shivered rocks, where grimy imps were at work blasting for ore, a string of adventurers laden with picks, shovels, and crowbars; kegs of powder, frying-pans, pitch-forks, and other instruments of torture—all wearily toiling in the same direction; decrepit old men, with avarice imprinted upon their furrowed brows; Jews and Gentiles, foot-weary and haggard; the young and the old, the strong and the weak, all alike burning with an unhallowed lust for lucre; and then I shuddered as the truth flashed upon me that they were going straight to—Virginia City.

Every foot of the cañon was claimed, and gangs of miners were at work all along the road, digging and delving into the earth like so many infatuated gophers. Many of these unfortunate creatures lived in holes dug into the side of the hill, and here and there a blanket thrown over a few stakes served as a domicile to shield them from the weather.

At Gold Hill, two miles beyond the Gate, the excitement was quite pitiable to behold. Those who were not at work burrowing holes into the mountain were gathered in gangs around the whisky saloons, pouring liquid fire down their throats, and swearing all the time in a manner so utterly reckless as to satisfy me they had long since bid farewell to hope.

THE DEVIL'S GATE.

This district is said to be exceedingly rich in gold, and I fancy it may well be so, for it is certainly rich in nothing else. A more barren-looking and forbidding spot could scarcely be found elsewhere on the face of the earth. The whole aspect of the country indicates that it must have been burned up in hot fires many years ago and reduced to a mass of cinders, or scraped up from all the desolate spots in the known world, and thrown over the Sierra Nevada Mountains in a confused mass to be out of the way. I do not wish to be understood as speaking disrespectfully of any of the works of creation, but it is inconceivable that this region should ever have been designed as an abode for man.

A short distance beyond Gold Hill we came in sight of the great mining capital of Washoe, the far-famed Virginia City. In the course of a varied existence it had been my fortune to visit the city of Jerusalem, the city of Constantinople, the city of the Sea, the City of the Dead, the Seven Cities, and others of historical celebrity in the Old World, and many famous cities in the New, including Port Townsend, Crescent City, Benicia, and the New York of the Pacific, but I had never yet beheld such a city as that which now burst upon my distended organs of vision.

On a slope of mountains speckled with snow, sage-bushes, and mounds of upturned earth, without any apparent beginning or end, congruity or regard for the eternal fitness of things, lay outspread the wondrous city of Virginia.

Frame shanties, pitched together as if by accident; tents of canvas, of blankets, of brush, of potato-sacks and old shirts, with empty whisky-barrels for chimneys; smoky hovels of mud and stone; coyote holes in the mountain side forcibly seized and held by men; pits and shafts with smoke issuing from every crevice; piles of goods and rubbish on craggy points, in the hollows, on the rocks, in the mud, in the snow, every where, scattered broadcast in pell-mell confusion, as if the clouds had suddenly burst overhead and rained down the dregs of all the flimsy, rickety, filthy little hovels and rubbish of merchandise that had ever undergone the process of evaporation from the earth since the days of Noah. The intervals of space, which may or may not have been streets, were dotted over with human beings of such

VIRGINIA CITY.

sort, variety, and numbers, that the famous ant-hills of Africa were as nothing in the comparison. To say that they were rough, muddy, unkempt and unwashed, would be but faintly expressive of their actual appearance; they were all this by reason of exposure to the weather; but they seemed to have caught the very diabolical tint and grime of the whole place. Here and there, to be sure, a San Francisco dandy of the "boiled shirt" and "stove-pipe" pattern loomed up in proud consciousness of the triumphs of art under adverse circumstances, but they were merely peacocks in the barn-yard.

A fraction of the crowd, as we entered the precincts of the town, were engaged in a lawsuit relative to a question of title. The arguments used on both sides were empty whisky-bottles, after the fashion of the *Basilinum*, or club law, which, according to Addison, prevailed in the colleges of learned men in former times. Several of the disputants had already been knocked down and convinced, and various others were freely shedding their blood in the cause of justice. Even the bull-terriers took an active part—or, at least, a very prominent part. The difficulty was about the ownership of a lot, which had been staked out by one party and "jumped" by another. Some two or three hundred disinterested observers stood by, enjoying the spectacle, several of them with their hands on their revolvers, to be ready in case of any serious issue; but these dangerous weapons are only used on great occasions—a refusal to drink, or some illegitimate trick at monte.

Upon fairly reaching what might be considered the centre of the town, it was interesting to observe the manners and customs of the place. Groups of keen speculators were huddled around the corners, in earnest consultation about the rise and fall of stocks; rough customers, with red and blue flannel shirts, were straggling in from the Flowery Diggings, the Desert, and other rich points, with specimens of croppings in their hands, or offering bargains in the "Rogers," the "Lady Bryant,"

A QUESTION OF TITLE.

the "Mammoth," the "Woolly Horse," and Heaven knows how many other valuable *leads*, at prices varying from ten to seventy-five dollars a foot. Small knots of the knowing ones were in confidential interchange of thought on the subject of every other man's business; here and there a loose man was caught by the button, and led aside behind a shanty to be "stuffed;" every body had some grand secret, which nobody else could find out; and the game of "dodge" and "pump" was universally played. Jew clothing-men were setting out their goods and chattels in front of wretched-looking tenements; monte-dealers, gamblers, thieves, cut-throats, and murderers were mingling miscellaneously in the dense crowds gathered around the bars of the drinking saloons. Now and then a half-starved Pah-Ute or Washoe Indian came tottering along under a heavy press of fagots and whisky. On the main street, where the mass of the population were gathered, a jaunty fellow who had "made a good thing of it" dashed through the crowds on horseback, accoutred in genuine Mexican style, swinging his *riata* over his head, and yelling like a devil let loose. All this time the wind blew in terrific gusts from the four quarters of the compass, tearing away signs, capsizing tents, scattering the grit from the gravel-banks with blinding force in every body's eyes, and sweeping furiously around every crook and corner in search of some sinner to smite. Never was such a wind as this—so scathing, so searching, so given to penetrate the very core of suffering humanity; disdaining overcoats, and utterly scornful of shawls and blankets. It actually seemed to double up, twist, pull, push, and screw the unfortunate biped till his muscles cracked and his bones rattled—following him wherever he sought refuge, pursuing him down the back of the neck, up the coat-sleeves, through the legs of his pantaloons, into his boots—in short, it was the most villainous and persecuting wind that ever blew, and I boldly protest that it did nobody good.

Yet, in the midst of the general wreck and crash of matter, the business of trading in claims, "bucking" and "bearing," went on as if the zephyrs of Virginia were as soft and balmy as those of San Francisco.

"MY CLAIM, SIR!"

This was surely— No matter; nothing on earth could aspire to competition with such a place. It was essentially infernal in every aspect, whether viewed from the Comstock Ledge or the summit of Gold Hill. Nobody seemed to own the lots except by right of possession; yet there was trading in lots to an unlimited extent. Nobody had any money, yet every body was a millionaire in silver claims. Nobody had any credit, yet every body bought thousands of feet of glittering ore. Sales were made in the Mammoth, the Lady Bryant, the

Sacramento, the Winnebunk, and the innumerable other "outside claims," at the most astounding figures, but not a dime passed hands. All was silver under ground, and deeds and mortgages on top; silver, silver every where, but scarce a dollar in coin. The small change had somehow gotten out of the hands of the public into the gambling saloons.

Every speck of ground covered by canvas, boards, baked mud, brush, or other architectural material, was jammed to suffocation; there were sleeping houses, twenty feet by thirty, in which from one hundred and fifty to two hundred solid sleepers sought slumber at night, at a dollar a head; tents, eight by ten, offering accommodations to the multitude; any thing or any place, even a stall in a stable, would have been a luxury.

The chief hotel, called, if I remember, the "Indication," or the "Hotel de Haystack," or some such euphonious name, professed to accommodate three hundred live men, and it doubtless did so, for the floors were covered from the attic to the solid earth—three hundred human beings in a tinder-box not bigger than a first-class hen-coop! But they were sorry-looking sleepers as they came forth each morning, swearing at the evil genius who had directed them to this miserable spot—every man a dollar and a pound of flesh poorer. I saw some, who perhaps were short of means, take surreptitious naps against the posts and walls in the bar-room, while they ostensibly professed to be mere spectators.

In truth, wherever I turned there was much to confirm the forebodings with which I had entered the Devil's Gate. The deep pits on the hill-sides; the blasted and barren appearance of the whole country; the unsightly hodge-podge of a town; the horrible confusion of tongues; the roaring, raving drunkards at the bar-rooms, swilling fiery liquids from morning till night; the flaring and flaunting gambling-saloons, filled with desperadoes of the vilest sort; the ceaseless torrent of imprecations that shocked the ear on every side; the mad specula-

GOLD HILL.

tions and feverish thirst for gain—all combined to give me a forcible impression of the unhallowed character of the place.

What dreadful savage is that? I asked, as a ferocious-looking monster in human shape stalked through the crowd. Is it—can it be the —? No; that's only a murderer. He shot three men a few weeks ago, and will probably shoot another before night. And this aged and decrepit man, his thin locks floating around his haggard and unshaved face, and matted with filth? That's

SAN FRANCISCO SPECULATORS.

a speculator from San Francisco. See how wildly he grasps at every "indication," as if he had a lease of life for a thousand years! And this bull-dog fellow, with a mutilated face, button-holing every by-passer? That fellow? Oh, he's only a "bummer" in search of a cocktail. And this—and this—all these crazy-looking wretches, running hither and thither with hammers and stones in their hands, calling one another aside, hurrying to the assay offices, pulling out papers, exchanging mysterious

ASSAY OFFICE.

signals—who and what are all these? Oh, these are Washoe millionaires. They are deep in "outside claims." The little fragments of rock they carry in their hands are "croppings" and "indications" from the "Wake-up-Jake," "Root-Hog-or-Die," "Wild-Cat," "Grizzly-Hill," "Dry-up," "Same Horse," "Let-her-Rip," "You Bet," "Gouge-Eye," and other famous ledges and companies, in which they own some thousands of feet. Hold, good friend! I am convinced there is no rest for the wicked. All night long these dreadful noises continue; the ears are distracted with an unintelligible jargon of "croppings," "ledges," "lodes," "leads," "indications," "feet," and "strikes," and the nostrils offended with foul odors of boots, old pipes, and dirty blankets—who can doubt the locality? If the climate is more rigorous than Dante describes it—if Calypso might search in vain for Ulysses in such a motley crowd—these apparent differences are not inconsistent with the general theory of changes produced by American emigration and the sudden conglomeration of such incongruous elements.

I was grieved and astonished to find many friends here—some of them gentlemen who had borne a very fair reputation in San Francisco, and whose unhappy fate I never could have anticipated. The bankers and brokers who had been cut off, after a prosperous career on Montgomery Street, had, of course, reached the goal toward which they had long been tending; the lawyers, who had set their unfortunate fellow-creatures by the ears, were now in a congenial element; the hard traders and unscrupulous speculators, who had violated all the moral obligations of life in their greedy lust for money, naturally abounded in large numbers; in short, it was not a matter of surprise that justice had at length been dealt out to many sinful men. But when I recognized friends whom I had formerly known as good citizens, the fathers of interesting families, exemplary members of society in San Francisco, I was profoundly shocked. It was impossible to deny that they must have been guilty of some grievous wickedness to entitle them to such a punishment.

What surprised me most of all was to find Colonel R——, to whom I had a letter of introduction, the leading spirit here. His assistance was sought by all. He was the best friend to any man in need of advice. Hospitality with him was a cardinal virtue. He had turned out of his own snug quarters long since to make room for the sick and disabled, and now slept about wherever he could find shelter. He was chief owner in the "Comstock Lead," and showed great liberality in giving a helping hand to others on the road to fortune. In fine, I am utterly unable to determine for what crime he was now suffering expiation. There was nothing in his conduct that I could discover the least unbecoming to a good citizen. His benevolence, hospitality, and genial manners were worthy any Christian. To me and to many others he proved the good Samaritan, and I still hesitate to believe that he merited the hard fate now meted out to him. But who can fathom the judgments pronounced upon men?

The bare contemplation of the miseries suffered by the inhabitants of this dreadful place was enough to stagger all convictions of my identity. Could it be possible that I was at last in—in Virginia City? What had I done to bring me to this? In vain I entered into a retrospection of the various iniquities of my life; but I could hit upon nothing that seemed bad enough to warrant such a fate. At length a withering truth flashed upon me. This must be the end of a federal existence! This must be the abode of ex-inspector-generals! It must be here that the accounts current of the decapitated are examined. Woe to the wretch who failed to profit by specie clause of the Independent Treasury Act while he had official claws on hand! Such *laches* of public duty can not be tolerated even in—Virginia City.

I slept, or rather tried to sleep, at one "Zip's," where there were only twenty "bunks" in the room, and was fortunate in securing a bunk even there. But the great Macbeth himself, laboring under the stings of an evil conscience, could have made a better hand of sleeping than I did at Zip's. It proved to be a general meeting-place for my San Francisco friends, and as they were all very rich in mining claims, and bent on getting still richer, they were continually making out deeds, examining

A FALL.

titles, trading and transferring claims, discussing the purchases and prospects of the day, and exhibiting the most extraordinary "indications" yet discovered, in which one or other of them held an interest of fifty or a hundred feet, worth, say, a thousand dollars a foot. Between the cat-naps of oblivion that visited my eyes there was a constant din of "croppings"—"feet"—"fifty thousand dollars"—"struck it rich!"—"the Comstock Ledge!"—"the Billy Choller!"—"Miller on the rise!"—"Mammoth!"—"Sacramento!"—"Lady Bryant!"—"a thousand feet more!"—"great bargain"—"forty dollars a foot!"—crash! rip! bang!—"an earthquake!"—"run for your lives!"

What the deuce is the matter?

It happened thus one night. The wind was blowing in terrific gusts. In the midst of the general clatter on the subject of croppings, bargains, and indications, down came our next neighbor's house on the top of us with a terrific crash. For a moment it was difficult to tell which house was the ruin. Amid projecting and shivered planks, the flapping of canvas, and the howling of the wind, it really seemed as if chaos had come again. But "Zip's" was well braced, and stood the shock without much damage, a slight heel and lurch to leeward being the chief result. I could not help thinking, as I turned in again after the alarm, that there could no longer be a doubt on the subject which had already occasioned me so many unpleasant reflections. It even seemed as if I smelled something like brimstone; but, upon calling to Zip to know what was the matter, he informed me that he was "only dryin' the boots on the stove."

CHAPTER V.

SOCIETY OF VIRGINIA CITY.

NOTWITHSTANDING the number of physicians who had already hoisted their "shingles," there was much sickness in Virginia, owing chiefly to exposure and dissipation, but in some measure to the deleterious quality of the water. Nothing more was wanting to confirm my original impressions. The water was certainly the worst ever used by man. Filtered through the Comstock Lead, it carried with it much of the plumbago, arsenic, copperas, and other poisonous minerals alleged to exist in that vein. The citizens of Virginia had discovered what they conceived to be an infallible way of "correct-

THE COMSTOCK LEAD.

ing it;" that is to say, it was their practice to mix a spoonful of water in half a tumbler of whisky, and then drink it. The whisky was supposed to neutralize the bad effects of the water. Sometimes it was considered good to mix it with gin. I was unable to see how any advantage could be gained in this way. The whisky contained strychnine, oil of tobacco, tarentula juice, and various effective poisons of the same general nature, including a dash of corrosive sublimate; and the gin was manufactured out of turpentine and whisky, with a sprinkling of prussic acid to give it flavor. For my part, I preferred taking poison in its least complicated form, and therefore adhered to the water. With hot saleratus bread, beans fried in grease, and such drink as this, it was no wonder that scores were taken down sick from day to day.

Sickness is bad enough at the best of times, but here the condition of the sick was truly pitiable. There was scarcely a tenement in the place that could be regarded as affording shelter against the piercing wind; and crowded as every tent and hovel was to its utmost capacity, it was hard even to find a vacant spot to lie down, much less sleep or rest in comfort. Many had come with barely means sufficient to defray their expenses to the diggings, in the confident belief that they would immediately strike upon "something rich;" or, if they failed in that, they could work a while on wages. But the highest wages here for common labor were three dollars a day, while meals were a dollar each, and lodgings the same. It was a favor to get work for "grub." Under such circumstances, when a poor fellow fell sick, his recovery could only be regarded as a matter of luck. No record of the deaths was kept. The mass of the emigration were strangers to each other, and it concerned nobody in particular when a man "pegged out," except to put him in a hole somewhere out of the way.

I soon felt the bad effects of the water. Possibly I had committed an error in not mixing it with the other poisons; but it was quite poisonous enough alone to give me violent pains in the stomach and a very severe diarrhea. At the same time, I was seized with an acute attack of rheumatism in the shoulder and neuralgic pains in the head. The complication of miseries which I now suffered was beyond all my calculations of the hardships of mining life. As yet I had struck nothing better than "Winn's Restaurant," where I took my meals. The Comstock Ledge was all very fine, but a THOUSAND DOLLARS A FOOT! Who ever had a thousand dollars to put in a running foot of ground, when not even the great Comstock himself could tell where it was running to. On the whole, I did not consider the prospect cheering.

At this period there were no laws of any kind in the district for the preservation of order. Some regulations had been established to secure the right of discovery to claimants, but they were loose and indefinite, differing in each district according to the caprice of the miners, and subject to no enforcement except that of the revolver. In some localities the original discoverer of a vein was entitled to 400 running feet; he could put down the names of as many friends as he chose at 200 feet each. Notice had to be recorded at certain places of record, designating the date and location of discovery. All "leads" were taken up with their dips, spurs, and angles." But who was to judge of the "dips, spurs, and angles?" That was the difficulty. Every man ran them to suit himself. The Comstock Ledge was in a mess of confusion. The shareholders had the most enlarged views of its "dips, spurs, and angles;" but those who struck croppings above and below were equally liberal in their notions; so that, in fine, every body's spurs were running into every body else's angles. The Cedar Hill Company were spurring the Miller Company; the Virginia Ledge was spurring the Continuation; the Dow Company were spurring the Billy Choller, and so on. It was a free fight all 'round, in which the dips, spurs, and angles might be represented thus, after the pattern of a bunch of snakes:

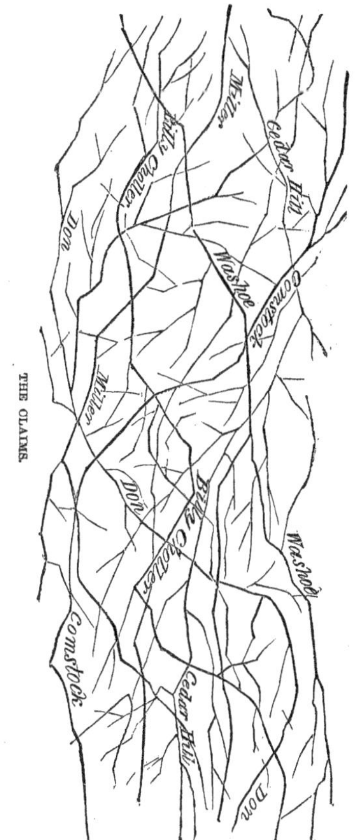

THE CLAIMS.

The contention was very lively. Great hopes were entertained that when Judge Cradlebaugh arrived he would hold court, and then there would be some hope of settling these conflicting claims. I must confess I did not share in the opinion that law would settle any dispute in which silver was concerned. The Almaden

Mine case is not yet settled, and never will be as long as there are judges and juries to sit upon it, and lawyers to argue it, and silver to pay expenses. Already Virginia City was infested with gentlemen of the bar, thirsting and hungering for chances at the Comstock. If it could only be brought into court, what a picking of bones there would be!

When the snow began to clear away there was no end to the discoveries alleged to be made every day. The Flowery Diggings, six miles below Virginia, were represented to be wonderfully rich—so rich, indeed, that the language of every speculator who held a claim there partook of the flowery character of the diggings. The whole country was staked off to the distance of twenty or thirty miles. Every hill-side was grubbed open, and even the Desert was pegged, like the sole of a boot, with stakes designating claims. Those who could not spare time to go out "prospecting" hired others, or furnished provisions and pack-mules, and went shares. If the prospecting party struck "any thing rich," it was expected they would share it honestly; but I always fancied they would find it more profitable to hold on to that, and find some other rich lead for the resident partners.

In Virginia City, a man who had been at work digging a cellar found rich indications. He immediately laid claim to a whole street covered with houses. The excitement produced by this "streak of luck" was perfectly frantic. Hundreds went to work grubbing up the ground under their own and their neighbors' tents, and it was not long before the whole city seemed in a fair way of being undermined. The famous *Winn*, as I was told, struck the richest lead of all directly under his restaurant, and was next day considered worth a million of dollars. The dips, spurs, and angles of these various discoveries covered every foot of ground within an area of six miles. It was utterly impossible that a fraction of the city could be left. Owners of lots protested in vain. The mining laws were paramount where there was no

"SILVER, CERTAIN, SIR!"

law at all. There was no security to personal property, or even to persons. He who turned in to sleep at night might find himself in a pit of silver by morning. At least it was thus when I made up my mind to escape from that delectable region; and now, four months later, I really don't know whether the great City of Virginia is still in existence, or whether the inhabitants have not found a "deeper deep, still threatening to devour."

It must not be supposed, from the general character of the population, that Virginia City was altogether destitute of men skilled in scientific pursuits. There were few, indeed, who did not profess to know something of geology; and as for assayers and assay offices, they were almost as numerous as barkeepers and groggeries. A tent, a furnace, half a dozen crucibles, a bottle of acid, and a hammer, generally comprised the entire establishment; but it is worthy of remark that the assays were always satisfactory. Silver, or indications of silver, were sure to be found in every specimen. I am confident some of these learned gentlemen in the assay business could have detected the precious metals in an Irish potato or a round of cheese for a reasonable consideration.

It was also a remarkable peculiarity of the country that the great "Comstock Lead" was discovered to exist in almost every locality, however remote or divergent from the original direction of the vein. I know a gentleman who certainly discovered a continuation of the Comstock forty miles from the Ophir mines, and at an angle of more than sixty degrees. But how could the enterprising adventurer fail to hit upon something rich, when every clod of earth and fragment of rock contained, according to the assays, both silver and gold? There was not a coyote hole in the ground that did not develop "indications." I heard of one lucky fellow who struck upon a rich vein, and organized an extensive company on the strength of having stumped his toe. Claims were even staked out and companies organized on "indications" rooted up by the squirrels and gophers. If they

"INDICATIONS, SURE!"

were not always indications of gold or silver, they were sure to contain copper, lead, or some other valuable mineral—plumbago or iridium, for instance. One man actually professed to have discovered "ambergris;" but I think he must have been an old whaler.

The complications of ills which had befallen me soon became so serious that I resolved to get away by hook or crook, if it was possible to cheat the —— corporate authorities of their dues. I had not come there to enlist in the service of Mammon at such wages.

Bundling up my pack one dark morning, I paid "Zip" the customary dollar, and while the evil powers were roistering about the grog-shops, taking their early bitters, made good my escape from the accursed place. Weak as I was, the hope of never seeing it again gave me nerve; and when I ascended the first elevation on the way to Gold Hill, and cast a look back over the confused mass of tents and hovels, and thought of all I had suffered there in the brief space of a few days, I involuntarily exclaimed, "If ever I put foot in that hole again, may the—"

But perhaps I had better not use strong language till I once more get clear of the Devil's Gate.

CHAPTER VI.

ESCAPE FROM VIRGINIA CITY.

As ill luck would have it, a perfect hurricane swept through the cañon from Gold Hill, sometimes in gusts so sudden and violent that it was utterly impossible to make an inch of headway. Tents were shivered and torn to shreds all along the wayside. I saw one party sitting at breakfast with nothing but the four posts which had originally sustained their tent and a few fragments of canvas flapping from them as a protection against the wind. Nothing could withstand its terrific force. Cabins with bush tops were unroofed; frame shanties were rent asunder, and the boards flew about like feathers; the air was filled with grit and drift, striking the face as if the great guns, which are sometimes said to blow, were loaded with duck-shot. Nor did the wind confine itself to one channel. It ranged up hill and down hill, raking the enemy fore and aft. In one place two tents were torn up, as one might say, by the roots, and carried off bodily to the top of the mountain; in another, half a dozen might be seen traveling down hill, at the rate of forty miles an hour, toward the Flowery Diggings. What became of all the unfortunate wretches who were thus summarily deprived of their local habitations I never learned. Most likely they sought refuge in the coyote holes, which, in fact, appeared to be untenanted; for I don't think coyotes could live long in such a country.

A short distance beyond Gold Hill a trail strikes off to the right, which is said to cut off four or five miles of the distance to Carson City. That would be a considerable gain to a traveler making his escape from Virginia City, and whose every step was attended with extreme physical suffering, to say nothing of the mental disquietude occasioned by his proximity to that place. Besides, it avoided the "Devil's Gate," of which I had also an intense dread. What hordes of dark and inexorable imps might be laying in wait there, with pitchforks to impale a poor fellow upon, and kegs of blasting powder to blow him up; what accounts might have to be rendered of one's stewardship at head-quarters; what particular kind of passport, sanded over with brimstone and stamped with a cloven-foot, might be demanded, it was not possible to conjecture. At all events, it was safer to incur no risk. The old adage of the "longest way round" did not occur to me.

I took the trail, and was soon out of sight of Gold City. The mountains were covered with snow, not very deep, but soft and slippery. In my weak state, with a racking rheumatism and the prostrating effects of the arsenic water, the labor of making headway against the fierce gusts of wind and keeping the trail was very severe. Every few hundred yards I had to lie down in the snow and await some relief from the paroxysms of pain. After an hour or two I reached a labyrinth of hills, in which the trail became lost by the melting of the snow. I still had some idea of the general direction, and kept on. My progress, however, was very slow, and at times so difficult that it required considerable effort of mind to avoid stopping altogether, and "taking the chances," as they say, in this agreeable region. Now all this may seem very absurd, as compared with the sufferings endured by Colonel Frémont in the Rocky Mountains, and doubtless is, in some respects. As, for instance: I was not shut up in a gorge of the mountains, a thousand miles from the habitations of man; I was not in a state of starvation, though thin enough for a starved man in all conscience; I was not at all likely to remain in any one position, however isolated, without being "spotted" by some enterprising miner in search of indications. But then, on the other hand, I was thoroughly dredged with arsenic, plumbago, copperas, and corrosive sublimate, and had neither mule nor "burro"—not even a woolly horse to carry me. Does any body pretend to say that the renowned arctic explorers ever encountered such a series of hardships as this? Four or five months of perpetual night, with the thermometer 80° below zero, may be uncomfortable; but then the adventurer in the polar regions has the advantage of being the farthest possible distance from certain other regions—say, from Virginia City.

About noon I came to the conclusion that, however willing the spirit might be, the flesh had done its best, and was now quite used up; so I stretched myself on the snow under a cedar bush, and resolved to await what assistance Providence might send me. I was not long there when a voice in the distance caught my ear. I rose and called. In a few minutes a mysterious figure emerged from the bushes at the mouth of a cañon a few hundred feet below. I beckoned to him to come up. The singular appearance and actions of the man attracted my attention.

His face was nearly black with dirt, and his hair was long and shaggy. On his head he wore a tattered cap, tied around the chin with a blue cotton handkerchief. A tremendous blue nose, a pair of green goggles, and boots extending up to his hips, completed the oddity of his appearance. At first he approached me rapidly; but at the distance of about fifty yards he halted, as if uncertain what to do. He then put down his pack, and began to search for something in the pockets of his coat—a knife, perhaps, or a pistol. Could it be possible this fellow was a robber, who had descried me from the opposite mountain, and was now bent upon murder? If so, it would be as well to bring the matter to an issue at once. I was unarmed, having even lost my penknife by reason of a rent in my pocket. There were desperate characters in this wilderness, who would think nothing of killing a man for his money; and although I had only about forty dollars left, that fact could not possibly be known to this marauder. His appearance, to be sure, was not formidable; but then one should not be too hasty in judging by appearances. For all I knew he might be the—Old Gentleman himself on a tour of inspection from Virginia City.

"Hallo, friend!" said I, assuming a conciliatory tone, "where are you bound?"

Upon this he approached a little closer. I soon perceived that he was a German Jew, who had either lost his way or was prospecting for silver. As he drew near, he manifested some signs of trepidation, evidently being afraid I would rob him of his pack, in which there was probably some jewelry or old clothes. It is hardly necessary for me to say that I had no intention of robbing him. I had not come to that yet. There was no telling to what straits I might be reduced; but, as long as I had a dollar in my pocket, I was determined to avoid highway robbery. Besides, it was beyond my strength at this particular crisis; a fact which the Jew seemed to recognize, for he now approached confidently. His first exclamation, on reaching the spot where I stood, was,

"Dank Gott! Ish dis de trail?"

"Where are you bound?"

"To Carson. I pe going to Carson, and I pe losht for six hours. Mein Gott! It ish an awful country. You know the way?"

"Of course. You don't suppose I'd be here if I didn't know the way?"

"Dat is zo."

"Come on, friend; I'm going in that direction. But don't walk very fast—I'm sick."

"Zo? Was is de matter?"

"Poisoned."

"Mein Gott! mein Gott! Das is awful."

"Very—it makes a fellow so weak."

"Mein Gott! Did dey poison you for your money?" And here the Jew put his hands behind him to see if his pack was safe.

"Oh no, it was only the water—arsenic and copperas."

"Zo!"

This explanation apparently relieved him of a very unpleasant train of thought, for he now became quite lively and talkative. As we trudged along, chatting sociably on various matters of common interest, it occurred to me from time to time that I had seen this man's face before. The idea grew upon me. It was not a matter of particular importance, and yet I could not banish it. His voice, too, was familiar. Certainly there was something about him that possessed an uncommon interest.

"Friend," said I, "it occurs to me I've seen you before."

"Zo? I dink de same."

Some moments elapsed before I could fix upon the oc-

AN OLD FRIEND.

casion or the place. All at once the truth flashed upon me. It was Strawberry Flat! I had slept with the man! This was the identical wretch who had robbed me of my stockings! In the excitement produced by the discovery and the recollection of my blistered feet, I verily believe, had I been armed with a broad-sword or battle-axe, after the fashion of Brian de Bois Guilbert, I would have cloven him in twain.

"Ha! I remember; it was at Strawberry! You slept with me one night," said I, in a tone of suppressed passion.

"Das is it! Das is it!" cried the Jew. "I shlept mit you at Sthrawberry!"

The effrontery of the villain was remarkable. Probably he would even acknowledge the theft.

"Friend," said I, calmly and deliberately, "did you miss a pair of woolen stockings in the morning about the time you started?"

"Look here!" quoth the wretch, suddenly halting, "was dey yours?"

"They were!"

At this the abominable rascal doubled himself up as if in a convulsion, shook all over, and turned almost black in the face. It was his mode of laughing.

"Well, I daught dey wos yours! I daught to myself, Mein Gott! how dat fellow will shwear when he find his sthockings gone!"

And here the convulsions were so violent that he fairly rolled over in the snow, and kicked as if in the agonies of death. It was doubtless very funny to rob a man of his valuable property and cause him days of suffering from blistered feet; but I was unable to see any wit in it till the Jew regained his breath and said,

"Vel, vel! I must sthand dhreat for dat! I know'd you'd shwear when you missed 'em. Vel, vel! das is goot! Here's a flask of first-rate brandy—dhrink!"

I took a small pull—medicinally, of course. From that moment my forgiveness was complete. I harbored not a particle of resentment against the man, though I never again could have entertained implicit confidence in his integrity.

In due time we reached the banks of Carson River at a place called Dutch John's, distant about four miles from Carson City. I have an impression that John was an emigrant from Salt Lake. He had brought with him a woman to whom he was "sealed," and was the father of a thriving little family of "cotton-heads." Some of the stage-drivers who were in the habit of taking a "smile" at John's persuaded him that he was now among a moral and civilized people, and must get married. To be "sealed" to a woman was not enough. He must be spliced according to Church and State, otherwise he would wake up some fine morning and find himself hanging to a tree. John had heard that the Californians were terrible fellows, and had a mortal dread of Vigilance Committees. The stage-drivers were rather a clever set of fellows, and no way strict in morals; but then they might hang him for fun, and what would be fun to them would be death to him. There was some charm in living an immoral life, to be sure, yet it would not do to enjoy that disreputable course at the expense of a disjointed neck. On the whole, John took the advice of the stage-drivers, and got married. Next day he rode through the streets of Carson, boasting of the adroit manner in which he had escaped the vengeance of the Vigilance Committee. I am happy to add that he is now a respectable member of the community. Not that I recommend his whisky. I consider it infinitely worse than any ever manufactured out of tobacco-juice, Cayenne pepper, and whale-oil at Port Townsend, Washington Territory, where the next worst whisky in the world is used as the common beverage of the inhabitants.

Leaving John's we came to the plain. Here the sand was heavy, and the walking very monotonous and tiresome. This part of Carson Valley is a complete desert. Scarcely a blade of grass was to be seen. Shriveled sage-bushes scattered here and there over the sand were the only signs of vegetation. Even the rabbits and sage-hens had abandoned the country. All the open spaces resembled the precincts of a slaughter-house. Cattle lay dead in every direction, their skulls, horns, and carcasses giving an exceedingly desolate aspect to the scene.

Near the river it was a perfect mass of corruption. Hundreds upon hundreds of bleached skeletons and rotting carcasses dotted the banks or lay in great mounds, where they had gathered for mutual warmth, and dropped down from sheer starvation. The smell filled the air for miles. Thousands of buzzards had gathered in from all parts to the great carnival of flesh—presenting a disgusting spectacle as they sat gorged and stupefied on the foul masses of carrion, they scarcely deigning to move as we passed. In the sloughs bordering on the river, oxen, cows, and horses were buried up to the necks where they had striven to get to the water, but, from excess of weakness, had failed to get back to the solid earth. Some were dead, others were dying. Around the latter the buzzards were already hovering, scarcely awaiting the extinction of life before they plunged in their ravenous beaks and tore out the eyes from the sockets. On the dry plain many hundreds of cattle had fallen from absolute starvation. The winter had been terribly severe, and the prolonged snows had covered what little vegetation there was. Those of the settlers who had saved hay enough for their stock found it more profitable to sell it at $300 a ton and let the stock die. Horses, oxen, and cows shared the same fate. Many lingered out the winter on the few stunted shrubs to be found on the foot-hills, and died just as the grass began to appear. It was a hard country for animals of all kinds. Those that were retained for the transportation of goods were little better than living skeletons, yet the amount of labor put upon them was extraordinary. In Virginia City it was almost impossible to procure a grain of barley for love or money. Enormous prices were offered for any kind of horse-feed by men who had come over on good horses, and who wished to keep them alive. At the rate of five dollars a day it required but a short time for the best horse to "eat his head off." Hay was sold in little wisps of a few pounds at sixty cents a pound, barley at seventy-five cents, and but little to be

CARSON VALLEY.

had even at those extravagant rates. A friend of mine from San Francisco, who arrived on a favorite horse, could get nothing in the way of feed but bread, and he paid fifty cents a loaf for a few scanty loaves about the size of biscuits to keep the poor animal alive. It was truly pitiable to see fine horses starving to death. The severity of the weather and the want of shelter were terribly severe on animals of every kind. Good horses could scarcely be sold for a tenth part of their cost, though the distance across the mountain could be performed under ordinary circumstances in two days. But where all was rush and confusion there was little time to devote to the calls of humanity. Men were crazy after claims. Every body had his fortune to make in a few months. The business of jockeying had not grown into full vogue except among a few, who were always willing to sell at very high prices and buy at very low —a remarkable fact connected with dealers in horseflesh.

The walk across Carson Valley through the heavy sand had exhausted what little of my strength remained, and I was about to give up the ghost for the third time, when a wagoner from Salt Lake gave me a lift on his wagon and enabled me to reach the town. Here my excellent friend Van Winkle gave me another chance in his bunk, and in the course of a few days I was quite recruited.

CHAPTER VII.

MY WASHOE AGENCY.

The courteous reader who has followed me so far will doubtless be disappointed that I have given so little practical information about the mines. Touching that I can only say, as Macaulay said of Sir Horace Walpole, the constitution of my mind is such that whatever is great appears to me little, and whatever is little seems great. The serious pursuits of life I regard as a monstrous absurdity on the part of mankind, especially rooting in the ground for money. The Washoe min nothing more than squirrel-holes on a difference being that squirrels burrow in th cause they live there, and men because they wa somewhere else. I deny and repudiate the idea tha man really has any necessity for money. He only think he does—which is a most unaccountable error.

But then you may have some notion of going to Washoe yourself, just to try your luck. Good friend, let me advise you—don't go. Stay where you are. Devote the remainder of your life to your legitimate business, your wife, and your baby. Don't go to Washoe. If you have no money, or but little, you had better go to—any other place. It is no retreat for a poor man. The working of silver mines requires capital. A poor man can not make wages in Washoe. If you are rich and wish to speculate—a word in your ear.

HOLDING ON TO IT.

"The undersigned is prepared to sell at reasonable prices" [this I quote from one of my advertisements] "valuable claims in the following companies:

The Dead Broke,	The Fool Hardy,
The Rip Snorter,	The Ousel Owl,
The Love's Despair,	The Grab Game,
The Ragged End,	The Riff-Raff.

"The titles to all these claims are perfect, and the purchaser of any claim will have no difficulty whatever in holding on to it."

I hope it will not be inferred from the desponding tone of my narrative that I deny the existence of silver in Washoe, for certainly nothing is farther from my intention. That there is silver in the Comstock Lead, and in great quantities, is a well-established fact. How many thousands of tons may be there it is impossible for me to say, but there must be an immense quantity—beyond all calculation in fact, as the ore is scattered all around the mines in great heaps, and every heap is said to be worth a fortune if it would only bear transportation to San Francisco at an expense of $600 per ton. The best of it is sorted out and packed off on mules every day or two, partly to get the silver out of it, and partly to show the speculators in San Francisco that the mines have not yet given out. The yield per ton is estimated at from $1200 to $2500. During the time of my visit to the mines but little work could be done on account of the number of speculators who were engaged in trying to sell out, few of them being disposed to engage in the slow operation of mining. Some said it was on account of the weather, but I suspect the weather had very little to do with it. The following is a rough estimate of the companies who claim to hold in the Comstock vein:

Billy Choller	1820 feet.	California	250 feet.
Hill and Norcross	250 "	Welch and Bryan	50 "
Goold and Curry	300 "	Central (again)	150 "
Savage	800 "	Ophir	200 "
Washoe	1200 "	Mexican	100 "
Belcher and Best	223 "	Continuation of Ophir	1200 "
Sides Ground	500 "	Newman, Scott, & Co.	300 "
Murphy	100 "	Miller Co.	3000 "
Kinney	60 "	Bob Allen and others	900 "
Central	100 "		

MOUNT OPHIR.

Besides about forty miles of outside claims, said to be on a direct line with the Comstock, and to be richer, if any thing, than the original vein.

When I left, the prices asked for a share in any of the above companies ranged from $200 to $2000 per running foot, and it was alleged that the purchaser could follow his running foot through all its dips, spurs, and angles. Some of these companies numbered as high as two or three hundred. I know a gentleman who sold out all his assets and invested the proceeds, $800, in 8 inches of the Central, and another who mortgaged his property to secure five feet in the Billy Choller. These gentlemen are, in all probability, at this moment worth a million of dollars each.

In short, the whole country looks black, blue, and white with silver, and where there is no silver there are croppings which indicate sulphurets or copperas.

CROPPINGS.

The Flowery Diggings were in full flower; and if they have since failed to realize the expectations that were then formed of them, it must be because the Mammoth lead gave out, or Lady Bryant did not sustain her reputation.

To the honest miner I have a word to say. You are

THE FLOWERY DIGGINGS.

S

HONEST MINER.

a free-born American citizen—that is, unless you were born in Ireland, which is so much the better, or in Germany, which is better still. You live by the sweat of your brow. You are God's noblest work—an honest man. The free exercise of the right of suffrage is guaranteed to you by the glorious Constitution of our common country. Upon your vote may depend the fate of millions of American freemen, nay, fate of Freedom itself, and the ultimate destiny of mankind. I do not appeal to you on the present occasion for any personal favor. Thank Fortune, I am beyond that. But in the name of common sense, in the name of our beloved state, in the name of the great Continental Congress, I do appeal to you, if you have a claim in California, HOLD ON TO IT! Don't go pirouetting about the country in search of better claims, abandoning ills that you are well acquainted with, and flying to others that you know nothing about. If you do, you may find it "a gloomy prospect."

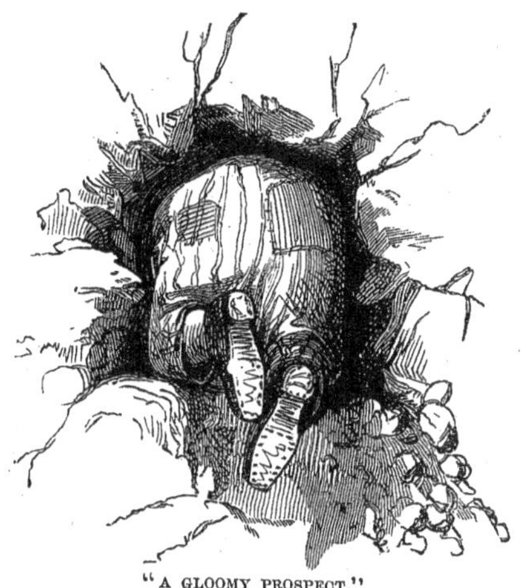

"A GLOOMY PROSPECT."

I was now, so to say, permanently established at Carson City. In other words, it was questionable whether I should ever be able to get away without resorting to the intervention of friends, which was an alternative too revolting for human nature to bear. The only resource left was "The Agency." I had forgotten all about it hithorto, and now resolved to call at the Express office, and see what fortune might be in store for me. Surely the advertisement must have elicited various orders of a lucrative nature. Nor was I disappointed. A package of letters awaited me. Without violating any confidential obligations, I may say, in general terms, that the contents and my answers were pretty much as follows:

A. Wishes to know what the prospect would be in Washoe for a young man of the medical profession. Has a small stock of drugs, and proposes to engage in the practice of medicine, and at the same time keep a drug store.

Ans. Doctors are already a drug in Washoe. Brandy, whisky, and gin are the only medicines taken. Bring over a lot of good liquors, prescribe them at two bits a dose, and you will do well. Charge, $10—please remit.

B. Has about twenty head of fine American cows. Would like to sell them, and wishes a contract made in advance.

Ans. Could find nobody who wanted to pay cash for cows. Money is scarce and cows are plenty. Have sold your cows, however, for the following valuable claims: 25 feet in the Root-Hog-or-Die; 40 feet in the Let-her-Rip; 50 feet in the Gone Case; and 100 feet in the You Bet. Charge, $25, which please remit by Express.

C. Would like to know if a school could be established in Washoe with any reasonable prospect of success. Has been engaged in the business for some years, and is qualified to teach the ordinary branches of a good English education, or, if desired, Greek and Latin.

Ans. No time to waste in learning here, and no use for the English language, much less Greek or Latin. A pious missionary might find occupation. One accustomed to mining could develop what indications there are of a spiritual nature among the honest miners. No charge.

D. Wishes to invest about $1500 in some good claims. Has three or four friends who will go in with him. Is willing to honor a draft for that amount. Hopes I will strike something rich.

Ans. Have bought a thousand feet for you in the very best silver mines yet discovered. They are all in and about the Devil's Gate. Several of them are supposed to be in the Comstock Ledge. They are worth $50,000 this moment; but if you can sell them in S. F. for an advance of $2000, do so by all means, as the silver may give out. Charge, $400 or nothing.

E. Has been in bad health for some time, and thinks a trip across the mountains would do him good. Please give him some information about the road and manner of living. How about lodgings and fare? Is troubled with the bronchitis, and wishes to know how the climate would be likely to affect it.

Ans. Hire a mule at Placerville, and if you are not too far gone the trip may benefit your bronchial tubes. The road is five feet deep by 130 miles long, and is composed chiefly of mountains, snow, and mud. Lodgings—from one to two hundred lodgers in each room, and from two to four bedfellows in each bed. Will not be troubled long with the bronchitis. The water will probably make an end of you in about two weeks. Charge—nothing.

F. Is a lawyer by profession, and desires to establish a business in some new country. Thinks there will be some litigation at Washoe in connection with the mines. Wishes to be informed on that point, and would be obliged for any general information.

Ans. About every tenth man in Washoe is a lawyer. There will doubtless be abundance of litigation there before long. Would advise you to go to some other new country, say Pike's Peak, for instance. Respecting things generally, Miller and Rodgers are going up and whisky down. Charge, 50 cents. Please remit.

G. Thinks of taking his family over to Washoe. How are the accommodations for women and children? And can servants be had?

Ans. Keep on thinking about that or something else, but don't attempt to carry your thoughts into effect. If you do, your wife must wear the—excuse me—she must wear male apparel. For accommodations, yourself and family might possibly be able to hire one bunk two feet by six; and you might seduce a Digger Indian to remain in your domestic employ by giving him $2 in cash and a gallon of whisky per day. Charge—nothing.

H. Has a house and lot worth about $10,000. Would like to trade it for some good mining claims. Can not sell the property for cash on account of a difficulty about the title; but this you need not mention, as it can probably be adjusted for a reasonable consideration.

Ans. Have traded your house and lot for 100 feet in the Pine Nut, 50 do. in the Ousel Owl, 50 do. in the Salmon Tail, 25 in the Roaring Jack, and 25 in the Amador. These are all good claims, and it will make no difference about the title to your house and lot, as each claim in the above-mentioned companies has also several titles to it. Charge, $500. Please remit.

I. Is in the stove business, and understands that cast-iron stoves bring a high price in Washoe. Has some notion of sending over a consignment. Please state expenses and prospect of success.

Ans. Stoves are very valuable in Washoe, especially cooking-stoves. It costs from 25 to 50 cents per pound to get them over on mule-back, at which prices they can be sold for claims, but not for money. If you have any very young stoves that can be planted, as the Schildbergers planted the salt, a good crop of them can be sold. Charge—nothing.

J. Is inventor of a process for extracting silver out of the crude ore without smelting. The machinery is simple, and would easily bear transportation. Could the patent right be sold in Washoe?

Ans. Nothing is more needed here than just such an invention as yours. Bring it over by all means. If you can extract silver out of the general average of the ore found here, either by smelting or otherwise, you will do a splendid business. Charge, $50. Please remit.

K. Understands that lumber is $300 a thousand in Virginia City. Can be delivered at the wharf in San Francisco from the Mendocino Mills for about $20 a thousand. Would it be practicable to get any quantity of it over, so as to make the speculation profitable?

Ans. You are correctly informed as to the value of lumber in Washoe. A balloon might be constructed to carry over a small lot; but, in case you found that mode of transportation too expensive, I know of no other way than to remove a portion of the Sierra Nevada Mountains in the rear of Placerville, or run a tunnel through underneath. It is possible that the price of labor might be an obstacle to the success of either of these plans, in which event, if you can contract to put one board on the back of each man leaving San Francisco, he may be able to earn his board, and you may be able to get your lumber over cheap. Charge, $25. Please remit.

I have thus given an average specimen of the letters that came pouring in upon me by every mail. It kept me busy, as may well be supposed, to attend to the numerous requests made by my correspondents; but the trouble was, no money came. There was a great deal, to be sure, for future collection, and as long as that was due it could not be lost by any injudicious speculation. It was some consolation, therefore, to reflect upon the large amount of capital that had accrued in the various operations of the Agency.

At this crisis, when fortune had fairly begun to smile, the weather changed again, and for days it stormed and snowed incessantly, covering up the whole valley, and blocking up every trail. A relapse of rheumatism and my poison-malady now seized me with renewed virulence. I had scarcely any rest by night or day, and soon saw that to remain would be a sure way of securing a claim to at least six feet of ground in the vicinity of Carson. The extraordinary number of persons who had invested in silver mines, and who were anxious to sell out in San Francisco, suggested the idea of changing my Agency to that locality. I therefore notified the public that there was a rare opportunity of selling out their claims to the best advantage, and it was not long before I was freighted down with "indications," powers of attorney, deeds, and bills of sale.

CHAPTER VIII.

START FOR HOME.

As soon as the weather permitted I set forth on my journey homeward, taking the stage to Genoa, in the hope of finding a horse or mule there upon which to cross the mountains. It was doubtful whether the trail was yet open; but a thaw had set in, and the prospect was that it would be practicable to get over in a few days. The stage from Genoa to Woodford's had been discontinued, in consequence of the expense of feeding the horses. All the saddle trains had left before the late snow, and there was not an animal of any kind to be had except by purchase—an alternation for which I was not prepared.

In this unfortunate state of affairs there was nothing left but to try it again on foot. It was with great difficulty that I could walk at all, much less carry my blankets and the additional weight of a heavy bundle of "croppings." The prospect of remaining at Genoa, however, was too gloomy to be thought of. So I sold my blankets for a night's lodging, and set out the next morning for Woodford's. By dint of labor and perseverance I accomplished about eight miles that day. It was dark night when I reached a small farm-house on the road-side. Here a worthy couple lived, who gave me comfortable lodgings, and cooked up such a luxurious repast of broiled chicken, toast, and tea, that I determined, if practicable, to remain a day or two, in order to regain my strength for the trip across the mountain.

The kindness and hospitality of these excellent people had the desired effect. In two days I was ready to proceed. Fortunately, an ox-wagon was going to Wood-

RETURN FROM WASHOE.

ford's for lumber, and I contracted with the driver, a good-humored negro, to give me a lift there for the sum of fifty cents.

I had the pleasure of meeting several San Francisco friends on the road, and gave them agreeable tidings of the mines. The trail had just been opened. A perfect torrent of adventurers came pouring over, forming an almost unbroken line all the way from Placerville. By this time the spring was well advanced and the excitement was at its height. The news from below was, that the whole state would soon be depopulated. Every body was coming—women, children, and all. Of course I wished them luck, but it was a marvel to me what they would do when they reached Washoe. Already there were eight or ten thousand people there, and not one in fifty had any thing to do, or could get employment for board and lodging. Companies were leaving every day for More's Lake and Walker's River, and the probability was that there would be considerable distress, if not absolute suffering. But it was useless to talk. Every adventurer must have a look at the diggings for himself. There must be luck in store for him, if for nobody else. For my part, I had taken a look and was satisfied.

The ox-team traveled very slowly, so that there was a good opportunity of seeing people pass both ways. The difference in the expression of the incoming and the outgoing was very remarkable, being about the difference between a man with fifty dollars in his pocket and one who wished to borrow that amount. There was that canny air of confidence about the former which betokens the possession of some knowledge touching the philosopher's stone not shared by mankind generally. About the latter there was a mingled expression of sadness and sarcasm, as if they were rather inclined to the opinion that some people had not yet seen the elephant.

As my ox carriage crept along uneasily over the rocky road, I was hailed from behind, "Hello dare! Sthop!" It was my friend the Jew again! I had lost sight of

OUTGOING AND INCOMING.

him in Carson, and now, by some fatality, he was destined to be my companion again.

"Mein Gott! I'm tired valking. Can't you give me a lift?" The driver was willing provided I had no objection. Now I had freely forgiven this man for the robbery of my stockings. I was not uncharitable enough to refuse help to a tired wayfarer; yet I had a serious objection to his company under existing circumstances. His boots were nearly worn out, and mine had but recently been purchased in Carson. If this fellow could embezzle my stockings and afterward unblushingly confess the act, what security could I have on the journey for the safety of my boots? I knew if he once started in with me he would never relinquish his claim to my company until we reached Placerville; for the fellow was rather of a sociable turn, and liked to talk. It seemed best, therefore, under all circumstances, to have a distinct understanding at once. The treaty was soon negotiated. On my part it was stipulated that Israel should ride to Woodford's on the ox wagon provided he paid his own fare; that we should cross the mountain together for mutual protection, provided he would deposit in my hands his watch or a $10 gold piece as security for the safety of my boots; and, finally, that he would bind himself by the most solemn obligations of honor not to steal both the security and the boots; to all of which the Jew assented with one of those internal convulsions which betokened great satisfaction in the arrangement. The watch was covered with pewter, as I discovered when he handed it to me; but I had no doubt it was worth eight or ten dollars. Besides, the treaty made no mention of the quality of the watch. It might possibly be an excellent timepiece, and, at all events, seemed to be worth a pair of boots.

Toward evening we arrived at Woodford's. Between two and three hundred travelers from the other side of the mountain had already gotten in, and it was represented that there was a line of pedestrians all the way over to Strawberry. The rush for supper was tremendous. Not even the famous Heenan and Sayers contest could compare with it, for here every body went in—or at least tried to get in. At the sixth round I succeeded in securing a favorable position, and when the battle commenced was fortunate enough to be crushed into a seat.

In the way of sleeping there was a general spread-out up stairs. By assuming a confidential tone with the proprietor I contrived to get a mattress and a pair of blankets. The Jew slept alongside on his pack, with a covering of loose coats. Nature's balmy restorer quickly put an end to all the troubles of the day, notwithstanding the incessant noise kept up throughout the night.

In the morning I awoke much refreshed. It was about seven o'clock, and time to start. I turned to arouse my friend Israel, but, to my surprise, found that he had already taken his departure. A horrible suspicion seized me. Had he also taken— Yes, of course; my boots were gone too! And the security? The watch? I looked under my pillow. Miserable wretch! he had also taken the watch. I might have known it! I was a fool for trusting him. When I picked up the

THE JEW'S BOOTS.

old pair of boots bequeathed to me as a token of remembrance by this depraved man—when I held them up to the light and examined them critically—when I reflected upon the journey before me, it was enough to bring tears to the sternest human eye.

No matter; I would catch the dastardly wretch on the trail. If ever I laid hands upon him again, so help me— But what is the use of swearing. No man ever caught another in this world with such a pair of boots on his feet—and here I examined them again—never! One might as well attempt to walk in a pair of condemned fire-buckets.

There was no help for it but to await some chance of getting over on horseback. Fortunately, a saddle-train which had passed down to Genoa during the previous day returned a little after daylight. For the sum of $30, cash in advance, I secured an unoccupied horse—the poorest animal, perhaps, ever ridden by mortal man. There is no good reason that I am aware of why people engaged in the horse-business should always select for my use the refuse of their stock; but such has invariably been their practice. I have never yet been favored with a horse that was not lame, halt, or blind, or otherwise physically afflicted.

I had not ridden more than a mile from Woodford's before I discovered that the miserable hack upon which I was mounted traveled diagonally, like a lugger beating against a head wind. His fore feet were well enough—they traveled on the trail; but his hind feet were continually undertaking to luff up a little to windward. When it is borne in mind that the trail was over a bank of snow from eight to ten feet deep, and not more than a foot wide, the inconvenience of that mode of locomotion will at once be perceived. Every few hundred yards the hind feet got off the trail, and went down with a sudden lurch that kept me in constant apprehension of being buried alive in the snow. Another serious difficulty was, that my horse, owing perhaps to the defect in his hind legs, had no capacity for short turns, so that whenever the trail suddenly diverged from its direct course, he invariably brought up against a rock, stump, or bank of snow.

I appealed to the captain or commander of the train to give me a better animal, but he assured me positively this was the very best in the whole lot, and that I would find him peculiarly adapted to mountain travel, where it was often an advantage for an animal to hold on to an upper trail with his fore feet while his hind ones were searching for another down below. In short, on this account solely he had named him "Guyascutas."

As there seemed to be no way of impressing the captain with a different opinion of the merits of Guyascutas, I was obliged to make the best of a bad bargain, and jog on as fast as spurs, blows, and entreaties could effect that result.

In reference to the Jew, whom I expected to overtake, and for whom I kept a sharp look-out, it may be as well to state at once that I never again put eyes on him. Whether he secreted himself behind some tree or rock till the saddle-train passed, or, overcome by remorse for the dastardly act he had committed, cast himself headlong over some precipice, I have never been able to ascertain. He is a miserable wretch at best. In view of the future, I would not for all the wealth of the Rothschilds stand in his— Well, yes, for that much money I might stand in his boots, provided no others were to be had; but I should regret extremely to be guilty of such an act toward any fellow-traveler as he had committed.

It was four o'clock when we got under way from the Lake House. A mule-driver from the other side of the divide had cautioned us against starting. There had been several snow-slides during the day, and it was only a few hours since the trail had been cut through. A large train of mules heavily laden must now be on the way down the Grade, and fifteen other trains had left Strawberry since noon.

SNOW SLIDE.

Those who have passed over the "Grade" can best appreciate our position. Two of our horses had already died of starvation and hard usage. There was no barley or feed of any kind to be had at the Lake House. The snow was rapidly melting, and avalanches might be expected at any moment. Only a day or two ago one of these fearful slides had occurred, sweeping all before it. Two mules and a horse were carried over the precipice and dashed to atoms, and the driver had barely escaped with his life.

It was considered perilous to stop on any part of the Grade. The trail was not over a foot wide, being heavily banked up on each side by the accumulated snow. Passing a pack train was very much like running a muck. The Spanish mules are so well aware of their privileges when laden, that they push on in defiance of all obstacles, often oversetting the unwary traveler by main force. I was struck with a barrel of whisky in one of the narrow passes some time previously and knocked nearly senseless, so that I had good cause to remember their prowess.

It was put to the vote whether we should make the attempt or remain, and finally, after much discussion, referred to our captain. He was evidently determined to go on at all hazards, having a stronger interest in the lives of his horses than any of the party.

At the word of command we mounted and put spurs to our jaded animals.

"Now, boys," said the captain, "keep together. Your lives depend upon it! Watch out for the pack trains, and when you see them coming hang on to a wide place! Don't come in contact with the pack-mules, or you'll go over the Grade certain."

There was no need of caution. Every nerve was strained to make the summit as soon as possible. It should be mentioned that the "Grade" is the Placerville state road, cut in the eastern slope of the Sierra Nevadas, and winding upward around each rib of the mountain for a distance of two miles. It was now washed away in many places by the melting of the snow, and some of the bridges across the ravines were in a very bad condition. From the first main elevation there is still another rise of two or three miles to the top of the divide, but this part is open and the ascent is comparatively easy. In meeting the pack trains the only hope of safety is to make for a point where the road widens. These places of security occur only three or four times in the entire ascent of the Grade. To be caught between them on a stubborn or unruly horse is almost certain destruction at this season of the year.

The only alternative is to dismount with all speed, wheel your horse round, and, if possible, get back to some place of security.

In about half an hour we made a point of rocks where the trail was bare. Our captain gave the order to dismount, and proceeded a short distance ahead to reconnoitre. The whole space occupied by our twelve horses and riders was not over six or eight feet wide by about thirty in length. Should any of the animals become stampeded, they were bound to go over. The tracks of several which had recently been pushed over the precipice by the pack trains were still visible. Our captain returned presently with news that a train was in sight. Soon we heard the tinkling of the bell attached to the leader, and then the clattering of the hoofs as the mules descended with their heavy burdens. One by one they passed. Whisky, gin, and brandy again! Barrels, half barrels, and kegs! The vaqueros made the cliffs resound with their Carambas and Carajas, their Doña Marias and Santa Sofias! a language apparently well understood by the mules. This was a train of forty mules, all laden with liquors for the thirsty miners. The vaqueros reported another train within half a mile of twenty-five mules, and others on the Grade.

After another train had passed, our captain gave the word to mount and "cut for our lives!" Scarcely five

THE GRADE.

seconds elapsed before we were all off, dashing helter-skelter up the trail. The horses plunged and stumbled over the rocks, slush, and mud in a manner truly pitiable for them and dangerous for us. In some places the mules had cut through for hundreds of yards, and the trail was perfectly honey-combed. But there was no time for humanity. Dashing the spurs into the bleeding sides of our animals, we pushed on as if all the evil powers of Virginia City were after us.

"Go it, boys!" our captain shouted; "neck or nothing! I see the train! Two hundred yards more and we're all safe! Caraja! Here's another train right on us!"

It was a palpable truth. The pack-mules came lumbering down around a point not fifty yards from us.

"Dismount all! Wheel! and cut back for your lives!" This was the order. In a moment we were all plunging frantically in the snow. Some of the horses were stampeded, and one man had gotten his riata around his leg. The mules had also commenced a stampede, when, by dint of shouting, plunging, and struggling, we got clear of them, and went tearing down the trail to our old station. The train soon passed us. Whisky again, of course. "How many trains more, senor?" to the vaquero. "Carambo! muchos! muchos! and on he went laughing. This was hard. We could not stand here much longer, for the tremendous bank of snow above us began to show indications of breaking away. Two trains more passed in rapid succession, and then our captain rode ahead again to reconnoitre. It was growing dusk. The prospect was any thing but cheering. At a given signal we mounted once more. Now commenced a terrible race. Heads, necks, legs, or horse-flesh were as nothing in the desperate struggle to reach the next point. This time we were in luck. The haven was attained just soon enough to avoid a train of forty mules. From the vaquero we learned that another was still on the Grade. We might be able to pass it, however, half a mile farther on. At the word of command we again mounted, and put spurs to our jaded animals. It was not long before we heard the tinkling of a bell. Now for it! halt! The mules were on us before we could turn; and here commenced a scene which baffles all description. Some of us were overturned, horses and all, in the banks of snow. Others sprang from their horses and let them struggle on their own account. All had to break a way out of the trail. The mules were stampeded, and kicked, brayed, and rolled by turns. The vaqueros were in a perfect frenzy of rage and terror combined — shrieking Maladetto! Carambo! and Caraja! till it seemed as if the reverberation must break loose the snow from above, and send an avalanche down on top of us all. Bridles got foul of stray legs and jerked the owners on their backs; riatas were twisted and wound around horses, mules, and whisky-barrels; packs went rolling hither and thither; men and animals kicked for their bare lives; heads, legs, and bodies were covered up in snow-drifts; and nobody knew what every body else was doing, or what he was doing himself. In short, the scene was altogether very lively, and would have been amusing had it not been intensified by the imminent risk of slipping over the precipice. It was at least a thousand feet down into Lake Valley, and a man might just as well be kicked on the head by twelve frantic horses and twenty-five vicious mules as undertake a trip down there by the short cut.

All troubles must end. Ours ended when the animals gave out for want of breath. Upon picking up our scattered regiment, with all arms and equipments used in the melée, we found the result as follows: Dead, none; wounded by kicks, scratches, sprains, and bruises, six; mortally frightened, the whole party, inclusive of our captain; lost, a keg of whisky, which some say went down to Lake Valley; but I have my suspicions where that keg went, and how it was secreted.

From this point over the summit we met several more

pack trains, and had an occasional tumble in the snow. Nothing more serious occurred. It was quite dark as we commenced our descent. The road here was a running stream of mud, obstructed by slippery rocks, ruts, stumps, and dead animals. It was a marvel to me how we ever reached the bottom without broken bones. My horse stumbled about every hundred yards, but never fell more than three quarters down. Somehow people rarely get killed in this country, unless shot by revolvers or bad whisky.

CHAPTER IX.

ARRIVAL IN SAN FRANCISCO.

The crowds were thicker than ever at Strawberry. From all accounts the excitement had only just commenced. Five thousand were represented to be on the road from the various diggings throughout California. I had bargained for a bed, and was enjoying the idea of a good supper—the savory odor of which came through the cracks of the bar-room door—when our captain announced that he could get no feed for his animals, and we must ride on to "Dick's," fourteen miles more. This was pretty tough on a sick man. The ride since morning had been quite hard enough to try the strength and temper of a well man; but add fourteen miles to that, of a dark night and raining into the bargain, and the sum total is not agreeable. It was useless to remonstrate. The captain was inflexible. He could not see his horses starve. One was just giving his last kick, and three more were about to "go in." I might stay if I pleased, suggested the captain, but the horses must go on. As I had paid thirty dollars for the ride, and had barely enough left to get to San Francisco, there was no alternative but to mount. By this time three of the party were so ill as to be scarcely able to sit in their saddles.

It is wonderful how much one can endure when there is nobody at hand to care a pin whether he lives or dies. I rather incline to the opinion that many people in this world die from the kindness and sympathy of friends, who, if thrown upon their own resources, would weather it out.

I have an impressive recollection of the fourteen miles from Strawberry to "Dick's." My horse, Guyascutas, broke down about half way. The rest of the party pushed on. About the same time the old torture of rheumatism and neuralgia assailed me in full force. It was pitch dark. There was no stopping-place nearer than "Dick's." The weather was cold, and a drenching rain had now penetrated my clothes to the skin.

A distinct recollection of my feelings a month ago, as I tramped along over this road with my pack on my back, afforded me ample material for philosophical reflection. Was it now somebody else—some decrepit old fogy who had lost his all, and had nothing more to expect in this world? Or could it possibly be the glowing enthusiast, just freed from the trammels of office, and inspired by visions of mountain life, liberty, and wealth? If it was the same—and there could hardly be any mistake about it, unless some mysterious translation of the spirit into some other body had taken place at Virginia Creek—the visions of mountain life, liberty, and unbounded riches were certainly of a very different character.

In addition to the peculiarity in the hind-quarters of Guyascutas, which caused him always to make two trails at the same time, I had now reason to suspect that he was entirely blind of one eye, and afflicted with a cataract on the other. Every hundred yards or so he walked off the road, and brought up in some deep cavity or against a pile of rocks. The mud in many places was up to his haunches, and if there was a comparatively dry spot any where in existence, he was sure to avoid it. I think he disliked me on account of the spurring I gave him on the Grade, and wanted to get rid of me in some way; or perhaps he considered his own course of life beyond farther endurance.

The result of all the stumbling, and running into deep pits, banks of rock, and mud-holes was, that I had to get down and walk the remainder of the way. If a conviction had not taken possession of my mind that the captain would compel me to pay for the horse in the event of failure to produce him, I would cheerfully have left him to his fate and proceeded alone; but, under the circumstances, I thought it best to lead him. At last the welcome lights hove in sight. It was not long before I was snugly housed at Dick's, where a good cup of tea brought life and hope back again. This, I may safely say, was my hardest day's experience of travel in any country.

Next day poor Guyascutas was so far gone on his long journey that I had to leave him at a stable on the roadside, and proceed on foot. By night I was within six miles of Placerville. Here I overtook a fellow-traveler, and bargained with him for his horse. From Placerville, by stage to Sacramento, the journey is devoid of interest. I arrived at San Francisco in due time, a little the worse for the wear, but still equal to any new emergency that might arise.

The citizens of San Francisco were on the *qui vive* for news from Washoe. Almost every man with a dollar to spare, and many who had nothing to spare, had invested, to a greater or less extent, in claims—from thousands of feet down to a few inches. Conflicting accounts had recently come down. The public mind was in a state of feverish excitement. Was Washoe a humbug, or was it not? Was there silver there, or was it all sham? What was the Ophir worth at this time? How about the Billy Choller and the Miller? These were but a few of the questions asked me on Montgomery Street. It required an hour to walk fifty yards, so great was the pressure for news. Could I tell any thing about the Winnemuck, or the Pine Nut, or the Rogers? Did I happen to know

RETURN TO SAN FRANCISCO.

T

what the Wake-up-Jake was worth in Washoe? What about the Lady Bryant—was it true that it had gone down? Whereabouts was the Jim Crack located, and what was Dead Broke worth? In short, I looked over more deeds, and answered more questions of a varied and indefinite nature, in the brief space of three days, than had ever been put to and answered by any one man before.

The editor of the *Bulletin*, who had made a flying visit to Washoe, and in whose company I had traveled down from Placerville, commenced about this time a series of articles, in which he told some startling truths. Base metal had been found in the Comstock; to what extent it prevailed nobody could tell. If the Comstock should prove to be worthless, what hope was there for the "outside claims."

The news spread like wild-fire. A panic seized upon the multitudes whose funds were invested in Washoe. Men hurried about the streets in search of purchasers of Washoe stock; but purchasers were nowhere to be found. Every body wanted to sell. The Comstock suddenly fell from one thousand down to five dollars per foot, and no sales at that. Miller went down fifty per cent.; and the Great Outside could scarcely be given away at any price! Alas! had it come to this? The gigantic Washoe speculation "gone in," and none so poor to do it reverence!

Softly! A word in your ear, reader! They are only "bucking it down" for purposes of speculation. The keen men who know a thing or two are buying up secretly. The silver is there, and it must come out. All this cry about base metal is "a dodge" to frighten the timid. If you have claims, hold on to them; they will be up again presently.

For my part, I thought it best to leave San Francisco before my correspondents—for whom, it will be remembered, I had executed some business in Washoe—retracted their good opinion of my sagacity. There was no chance at this crisis to sell the various claims with which I had been commissioned at Carson City. Capitalists were short of funds. The money-market was laboring under a depression. The liver of the body politic was in a state of collapse. I went to the principal bankers, but failed to accomplish any thing. They even refused to lend money on unquestionable security.

In view of all the circumstances, I determined to visit Europe. If the moneyed men of the Old World could only be satisfied of the extent, variety, and magnificence of the investments to be made in the New, they would not hesitate to open negotiations with an agent direct from Washoe.

Frankfort-on-the-Main, January, 1861.

You will perceive from my address, most esteemed reader, that I am now established at one of the best points for pecuniary transactions on the Continent of Europe. I have seen many of the wealthy burghers of Frankfort, and am pleased to say that they manifest a very friendly disposition. As yet they do not quite understand the nature of the proposed securities, but I have great confidence in their sagacity. My negotiations with the Rothschilds have been of the most amicable character. They have gone so far as to express the opinion that Washoe must be a remarkable country; and yesterday, when I proposed to sell them fifty feet in the Gone Case, and forty in the Roaring Grizzly, for the sum of one hundred thousand florins, they smiled so politely, and withal looked so completely puzzled, that I considered it best not to force an immediate answer. You are aware, of course, that in important negotiations of this kind it is judicious to let the opposite party sleep a night or two over your proposition. That the Rothschilds are at present a little wary of any investment in Washoe is quite natural. The nomenclature is new to them. They have never before heard of Roaring Grizzly and Gone Case silver mines. But if that should prove to be their only objection, I have no doubt they will ultimately purchase to the extent of several millions. If they do, I shall be happy to negotiate further sales for a reasonable commission, to be paid strictly in advance. My publishers will, I am confident, forward any letter to my address.

READING EXTRA BULLETIN.

THE END.

LOSSING'S FIELD-BOOK OF THE REVOLUTION. Pictorial Field-Book of the Revolution; or, Illustrations, by Pen and Pencil, of the History, Biography, Scenery, Relics, and Traditions of the War for Independence. By BENSON J. LOSSING. 2 vols., 8vo, Cloth, $14 00; Sheep, $15 00; Half Calf, $18 00; Full Turkey Morocco, $22 00.

LOSSING'S FIELD-BOOK OF THE WAR OF 1812. Pictorial Field-Book of the War of 1812; or, Illustrations, by Pen and Pencil, of the History, Biography, Scenery, Relics, and Traditions of the Last War for American Independence. By BENSON J. LOSSING. With several hundred Engravings on Wood, by Lossing and Barritt, chiefly from Original Sketches by the Author. 1088 pages, 8vo, Cloth, $7 00; Sheep, $8 50; Half Calf, $10 00.

WINCHELL'S SKETCHES OF CREATION. Sketches of Creation: a Popular View of some of the Grand Conclusions of the Sciences in reference to the History of Matter and of Life. Together with a Statement of the Intimations of Science respecting the Primordial Condition and the Ultimate Destiny of the Earth and the Solar System. By ALEXANDER WINCHELL, LL.D., Professor of Geology, Zoology, and Botany in the University of Michigan, and Director of the State Geological Survey. With Illustrations. 12mo, Cloth, $2 00.

WHITE'S MASSACRE OF ST. BARTHOLOMEW. The Massacre of St. Bartholomew: Preceded by a History of the Religious Wars in the Reign of Charles IX. By HENRY WHITE, M.A. With Illustrations. 8vo, Cloth, $1 75.

ALFORD'S GREEK TESTAMENT. The Greek Testament: with a critically-revised Text; a Digest of Various Readings; Marginal References to Verbal and Idiomatic Usage; Prolegomena; and a Critical and Exegetical Commentary. For the Use of Theological Students and Ministers. By HENRY ALFORD, D.D., Dean of Canterbury. Vol. I., containing the Four Gospels. 944 pages, 8vo, Cloth, $6 00; Sheep, $6 50.

ABBOTT'S HISTORY OF THE FRENCH REVOLUTION. The French Revolution of 1789, as viewed in the Light of Republican Institutions. By JOHN S. C. ABBOTT. With 100 Engravings. 8vo, Cloth, $5 00.

ABBOTT'S NAPOLEON BONAPARTE. The History of Napoleon Bonaparte. By JOHN S. C. ABBOTT. With Maps, Woodcuts, and Portraits on Steel. 2 vols., 8vo, Cloth, $10 00.

ABBOTT'S NAPOLEON AT ST. HELENA; or, Interesting Anecdotes and Remarkable Conversations of the Emperor during the Five and a Half Years of his Captivity. Collected from the Memorials of Las Casas, O'Meara, Montholon, Antommarchi, and others. By JOHN S. C. ABBOTT. With Illustrations. 8vo, Cloth, $5 00.

ADDISON'S COMPLETE WORKS. The Works of Joseph Addison, embracing the whole of the "Spectator." Complete in 3 vols., 8vo, Cloth, $6 00.

ALCOCK'S JAPAN. The Capital of the Tycoon: a Narrative of a Three Years' Residence in Japan. By Sir RUTHERFORD ALCOCK, K.C.B., Her Majesty's Envoy Extraordinary and Minister Plenipotentiary in Japan. With Maps and Engravings. 2 vols., 12mo, Cloth, $3 50.

ALISON'S HISTORY OF EUROPE. FIRST SERIES: From the Commencement of the French Revolution, in 1789, to the Restoration of the Bourbons, in 1815. [In addition to the Notes on Chapter LXXVI., which correct the errors of the original work concerning the United States, a copious Analytical Index has been appended to this American edition.] SECOND SERIES: From the Fall of Napoleon, in 1815, to the Accession of Louis Napoleon, in 1852. 8 vols., 8vo, Cloth, $16 00.

BANCROFT'S MISCELLANIES. Literary and Historical Miscellanies. By GEORGE BANCROFT. 8vo, Cloth, $3 00.

BALDWIN'S PRE-HISTORIC NATIONS. Pre-Historic Nations; or, Inquiries concerning some of the Great Peoples and Civilizations of Antiquity, and their Probable Relation to a still Older Civilization of the Ethiopians or Cushites of Arabia. By JOHN D. BALDWIN, Member of the American Oriental Society. 12mo, Cloth, $1 75.

BARTH'S NORTH AND CENTRAL AFRICA. Travels and Discoveries in North and Central Africa: being a Journal of an Expedition undertaken under the Auspices of H. B. M.'s Government, in the Years 1849–1855. By HENRY BARTH, Ph.D., D.C.L. Illustrated. 3 vols., 8vo, Cloth, $12 00.

HENRY WARD BEECHER'S SERMONS. Sermons by HENRY WARD BEECHER, Plymouth Church, Brooklyn. Selected from Published and Unpublished Discourses, and Revised by their Author. With Steel Portrait. Complete in 2 vols., 8vo, Cloth, $5 00.

LYMAN BEECHER'S AUTOBIOGRAPHY, &c. Autobiography, Correspondence, &c., of Lyman Beecher, D.D. Edited by his Son, CHARLES BEECHER. With Three Steel Portraits, and Engravings on Wood. In 2 vols., 12mo, Cloth, $5 00.

BELLOWS'S OLD WORLD. The Old World in its New Face: Impressions of Europe in 1867–1868. By HENRY W. BELLOWS. 2 vols., 12mo, Cloth, $3 50.

BOSWELL'S JOHNSON. The Life of Samuel Johnson, LL.D. Including a Journey to the Hebrides. By JAMES BOSWELL, Esq. A New Edition, with numerous Additions and Notes. By JOHN WILSON CROKER, LL.D., F.R.S. Portrait of Boswell. 2 vols., 8vo, Cloth, $4 00.

BRODHEAD'S HISTORY OF NEW YORK. History of the State of New York. By JOHN ROMEYN BRODHEAD. First Period, 1609–1664. 8vo, Cloth, $3 00.

BULWER'S PROSE WORKS. Miscellaneous Prose Works of Edward Bulwer, Lord Lytton. 2 vols., 12mo, Cloth, $3 50.

BURNS'S LIFE AND WORKS. The Life and Works of Robert Burns. Edited by ROBERT CHAMBERS. 4 vols., 12mo, Cloth, $6 00.

CARLYLE'S FREDERICK THE GREAT. History of Friedrich II., called Frederick the Great. By THOMAS CARLYLE. Portraits, Maps, Plans, &c. 6 vols., 12mo, Cloth, $12 00.

CARLYLE'S FRENCH REVOLUTION. History of the French Revolution. Newly Revised by the Author, with Index, &c. 2 vols., 12mo, Cloth, $3 50.

CARLYLE'S OLIVER CROMWELL. Letters and Speeches of Oliver Cromwell. With Elucidations and Connecting Narrative. 2 vols., 12mo, Cloth, $3 50.

CHALMERS'S POSTHUMOUS WORKS. The Posthumous Works of Dr. Chalmers. Edited by his Son-in-Law, Rev. WILLIAM HANNA, LL.D. Complete in 9 vols., 12mo, Cloth, $13 50.

COLERIDGE'S COMPLETE WORKS. The Complete Works of Samuel Taylor Coleridge. With an Introductory Essay upon his Philosophical and Theological Opinions. Edited by Professor SHEDD. Complete in Seven Vols. With a fine Portrait. Small 8vo, Cloth, $10 50.

CURTIS'S HISTORY OF THE CONSTITUTION. History of the Origin, Formation, and Adoption of the Constitution of the United States. By GEORGE TICKNOR CURTIS. 2 vols., 8vo, Cloth, $6 00.

DAVIS'S CARTHAGE. Carthage and her Remains: being an Account of the Excavations and Researches on the Site of the Phœnician Metropolis in Africa and other adjacent Places. Conducted under the Auspices of Her Majesty's Government. By Dr. DAVIS, F.R.G.S. Profusely Illustrated with Maps, Woodcuts, Chromo-Lithographs, &c. 8vo, Cloth, $4 00.

DRAPER'S CIVIL WAR. History of the American Civil War. By JOHN W. DRAPER, M.D., LL.D., Professor of Chemistry and Physiology in the University of New York. In Three Vols. 8vo, Cloth, $3 50 per vol.

DRAPER'S INTELLECTUAL DEVELOPMENT OF EUROPE. A History of the Intellectual Development of Europe. By JOHN W. DRAPER, M.D., LL.D., Professor of Chemistry and Physiology in the University of New York. 8vo, Cloth, $5 00.

DRAPER'S AMERICAN CIVIL POLICY. Thoughts on the Future Civil Policy of America. By JOHN W. DRAPER, M.D., LL.D., Professor of Chemistry and Physiology in the University of New York. Crown 8vo, Cloth, $2 50.

DU CHAILLU'S AFRICA. Explorations and Adventures in Equatorial Africa: with Accounts of the Manners and Customs of the People, and of the Chase of the Gorilla, the Crocodile, Leopard, Elephant, Hippopotamus, and other Animals. By PAUL B. DU CHAILLU. Numerous Illustrations. 8vo, Cloth, $5 00.

DU CHAILLU'S ASHANGO LAND. A Journey to Ashango Land: and Further Penetration into Equatorial Africa. By PAUL B. DU CHAILLU. New Edition. Handsomely Illustrated. 8vo, Cloth, $5 00.

DOOLITTLE'S CHINA. Social Life of the Chinese: with some Account of their Religious, Governmental, Educational, and Business Customs and Opinions. With special but not exclusive Reference to Fuhchau. By Rev. JUSTUS DOOLITTLE, Fourteen Years Member of the Fuhchau Mission of the American Board. Illustrated with more than 150 characteristic Engravings on Wood. 2 vols., 12mo, Cloth, $5 00.

EDGEWORTH'S (MISS) NOVELS. With Engravings. 10 vols., 12mo, Cloth, $15 00.

GIBBON'S ROME. History of the Decline and Fall of the Roman Empire. By EDWARD GIBBON. With Notes by Rev. H. H. MILMAN and M. GUIZOT. A new cheap Edition. To which is added a complete Index of the whole Work, and a Portrait of the Author. 6 vols., 12mo, Cloth, $9 00.

GROTE'S HISTORY OF GREECE. 12 vols., 12mo, Cloth, $18 00.

HALE'S (MRS.) WOMAN'S RECORD. Woman's Record; or, Biographical Sketches of all Distinguished Women, from the Creation to the Present Time. Arranged in Four Eras, with Selections from Female Writers of each Era. By Mrs. SARAH JOSEPHA HALE. Illustrated with more than 200 Portraits. 8vo, Cloth, $5 00.

HALL'S ARCTIC RESEARCHES. Arctic Researches and Life among the Esquimaux: being the Narrative of an Expedition in Search of Sir John Franklin, in the Years 1860, 1861, and 1862. By CHARLES FRANCIS HALL. With Maps and 100 Illustrations. The Illustrations are from Original Drawings by Charles Parsons, Henry L. Stephens, Solomon Eytinge, W. S. L. Jewett, and Granville Perkins, after Sketches by Captain Hall. 8vo, Cloth, $5 00.

HALLAM'S CONSTITUTIONAL HISTORY OF ENGLAND, from the Accession of Henry VII. to the Death of George II. 8vo, Cloth, $2 00.

HALLAM'S LITERATURE. Introduction to the Literature of Europe during the Fifteenth, Sixteenth, and Seventeenth Centuries. By HENRY HALLAM. 2 vols., 8vo, Cloth, $4 00.

HALLAM'S MIDDLE AGES. State of Europe during the Middle Ages. By HENRY HALLAM. 8vo, Cloth, $2 00.

HILDRETH'S HISTORY OF THE UNITED STATES. FIRST SERIES: From the First Settlement of the Country to the Adoption of the Federal Constitution. SECOND SERIES: From the Adoption of the Federal Constitution to the End of the Sixteenth Congress. 6 vols., 8vo, Cloth, $18 00.

HARPER'S PICTORIAL HISTORY OF THE REBELLION. Harper's Pictorial History of the Great Rebellion in the United States. With nearly 1000 Illustrations. In Two Vols., 4to. Price $6 00 per vol.

HARPER'S NEW CLASSICAL LIBRARY. Literal Translations.
The following Volumes are now ready. Portraits. 12mo, Cloth, $1 50 each.

CÆSAR.—VIRGIL.—SALLUST.—HORACE.—CICERO'S ORATIONS.—CICERO'S OFFICES, &c.—CICERO ON ORATORY AND ORATORS.—TACITUS (2 vols.).—TERENCE.—SOPHOCLES.—JUVENAL.—XENOPHON.—HOMER'S ILIAD.—HOMER'S ODYSSEY.—HERODOTUS.—DEMOSTHENES.—THUCYDIDES.—ÆSCHYLUS.—EURIPIDES (2 vols.).

HELPS'S SPANISH CONQUEST. The Spanish Conquest in America, and its Relation to the History of Slavery and to the Government of Colonies. By ARTHUR HELPS. 4 vols., 12mo, Cloth, $6 00.

HUME'S HISTORY OF ENGLAND. History of England, from the Invasion of Julius Cæsar to the Abdication of James II., 1688. By DAVID HUME. A new Edition, with the Author's last Corrections and Improvements. To which is Prefixed a short Account of his Life, written by Himself. With a Portrait of the Author. 6 vols., 12mo, Cloth, $9 00.

JAY'S WORKS. Complete Works of Rev. William Jay: comprising his Sermons, Family Discourses, Morning and Evening Exercises for every Day in the Year, Family Prayers, &c. Author's enlarged Edition, revised. 3 vols., 8vo, Cloth, $6 00.

JOHNSON'S COMPLETE WORKS. The Works of Samuel Johnson, LL.D. With an Essay on his Life and Genius, by ARTHUR MURPHY, Esq. Portrait of Johnson. 2 vols., 8vo, Cloth, $4 00.

KINGLAKE'S CRIMEAN WAR. The Invasion of the Crimea, and an Account of its Progress down to the Death of Lord Raglan. By ALEXANDER WILLIAM KINGLAKE. With Maps and Plans. Two Vols. ready. 12mo, Cloth, $2 00 per vol.

KRUMMACHER'S DAVID, KING OF ISRAEL. David, the King of Israel: a Portrait drawn from Bible History and the Book of Psalms. By FREDERICK WILLIAM KRUMMACHER, D.D., Author of "Elijah the Tishbite," &c. Translated under the express Sanction of the Author by the Rev. M. G. EASTON, M.A. With a Letter from Dr. Krummacher to his American Readers, and a Portrait. 12mo, Cloth, $1 75.

LAMB'S COMPLETE WORKS. The Works of Charles Lamb. Comprising his Letters, Poems, Essays of Elia, Essays upon Shakspeare, Hogarth, &c., and a Sketch of his Life, with the Final Memorials, by T. NOON TALFOURD. Portrait. 2 vols., 12mo, Cloth, $3 00.

LIVINGSTONE'S SOUTH AFRICA. Missionary Travels and Researches in South Africa; including a Sketch of Sixteen Years' Residence in the Interior of Africa, and a Journey from the Cape of Good Hope to Loando on the West Coast; thence across the Continent, down the River Zambesi, to the Eastern Ocean. By DAVID LIVINGSTONE, LL.D., D.C.L. With Portrait, Maps by Arrowsmith, and numerous Illustrations. 8vo, Cloth, $4 50.

LIVINGSTONE'S ZAMBESI. Narrative of an Expedition to the Zambesi and its Tributaries, and of the Discovery of the Lakes Shirwa and Nyassa. 1858-1864. By DAVID and CHARLES LIVINGSTONE. With Map and Illustrations. 8vo, Cloth, $5 00.

MARCY'S ARMY LIFE ON THE BORDER. Thirty Years of Army Life on the Border. Comprising Descriptions of the Indian Nomads of the Plains; Explorations of New Territory; a Trip across the Rocky Mountains in the Winter; Descriptions of the Habits of Different Animals found in the West, and the Methods of Hunting them; with Incidents in the Life of Different Frontier Men, &c., &c. By Brevet Brigadier-General R. B. MARCY, U.S.A., Author of "The Prairie Traveller." With numerous Illustrations. 8vo, Cloth, Beveled Edges, $3 00.

M'CLINTOCK & STRONG'S CYCLOPÆDIA. Cyclopædia of Biblical, Theological, and Ecclesiastical Literature. Prepared by the Rev. JOHN M'CLINTOCK, D.D., and JAMES STRONG, S.T.D. 3 *vols. now ready.* Royal 8vo. Price per vol., Cloth, $5 00; Sheep, $6 00; Half Morocco, $8 00.

MACAULAY'S HISTORY OF ENGLAND. The History of England from the Accession of James II. By THOMAS BABINGTON MACAULAY. With an Original Portrait of the Author. 5 vols., 8vo, Cloth, $10 00; 12mo, Cloth, $7 50.

MOSHEIM'S ECCLESIASTICAL HISTORY, Ancient and Modern; in which the Rise, Progress, and Variation of Church Power are considered in their Connection with the State of Learning and Philosophy, and the Political History of Europe during that Period. Translated, with Notes, &c., by A. MACLAINE, D.D. A new Edition, continued to 1826, by C. COOTE, LL.D. 2 vols., 8vo, Cloth, $4 00.

NEVIUS'S CHINA. China and the Chinese: a General Description of the Country and its Inhabitants; its Civilization and Form of Government; its Religious and Social Institutions; its Intercourse with other Nations; and its Present Condition and Prospects. By the Rev. JOHN L. NEVIUS, Ten Years a Missionary in China. With a Map and Illustrations. 12mo, Cloth, $1 75.

OLIN'S (DR.) LIFE AND LETTERS. 2 vols., 12mo, Cloth, $3 00.

OLIN'S (DR.) TRAVELS. Travels in Egypt, Arabia Petræa, and the Holy Land. Engravings. 2 vols., 8vo, Cloth, $3 00.

OLIN'S (DR.) WORKS. The Works of Stephen Olin, D.D., late President of the Wesleyan University. 2 vols., 12mo, Cloth, $3 00.

OLIPHANT'S CHINA AND JAPAN. Narrative of the Earl of Elgin's Mission to China and Japan, in the Years 1857, '58, '59. By LAURENCE OLIPHANT, Private Secretary to Lord Elgin. Illustrations. 8vo, Cloth, $3 50.

OLIPHANT'S (MRS.) LIFE OF EDWARD IRVING. The Life of Edward Irving, Minister of the National Scotch Church, London. Illustrated by his Journals and Correspondence. By Mrs. OLIPHANT. Portrait. 8vo, Cloth, $3 50.

PAGE'S LA PLATA. La Plata: the Argentine Confederation and Paraguay. Being a Narrative of the Exploration of the Tributaries of the River La Plata and Adjacent Countries, during the Years 1853, '54, '55, and '56, under the Orders of the United States Government. New Edition, containing Farther Explorations in La Plata, during 1859 and '60. By THOMAS J. PAGE, U.S.N., Commander of the Expeditions. With Map and numerous Engravings. 8vo, Cloth, $5 00.

POETS OF THE NINETEENTH CENTURY. The Poets of the Nineteenth Century. Selected and Edited by the Rev. ROBERT ARIS WILLMOTT. With English and American Additions, arranged by EVERT A. DUYCKINCK, Editor of "Cyclopædia of American Literature." Comprising Selections from the Greatest Authors of the Age. Superbly Illustrated with 132 Engravings from Designs by the most Eminent Artists. In elegant small 4to form, printed on Superfine Tinted Paper, richly bound in extra Cloth, Beveled, Gilt Edges, $6 00; Half Calf, $6 00; Full Turkey Morocco, $10 00.

PRIME'S COINS, MEDALS, AND SEALS. Coins, Medals, and Seals, Ancient and Modern. Illustrated and Described. With a Sketch of the History of Coins and Coinage, Instructions for Young Collectors, Tables of Comparative Rarity, Price-Lists of English and American Coins, Medals, and Tokens, &c., &c. Edited by W. C. PRIME, Author of "Boat Life in Egypt and Nubia," "Tent Life in the Holy Land," &c., &c. 8vo, Cloth, $3 50.

SPRING'S SERMONS. Pulpit Ministrations; or, Sabbath Readings. A Series of Discourses on Christian Doctrine and Duty. By Rev. GARDINER SPRING, D.D., Pastor of the Brick Presbyterian Church in the City of New York. Portrait on Steel. 2 vols., 8vo, Cloth, $6 00.

SHAKSPEARE. The Dramatic Works of William Shakspeare, with the Corrections and Illustrations of Dr. JOHNSON, G. STEEVENS, and others. Revised by ISAAC REED. Engravings. 6 vols., Royal 12mo, Cloth, $9 00.

SMILES'S LIFE OF THE STEPHENSONS. The Life of George Stephenson, and of his Son, Robert Stephenson; comprising, also, a History of the Invention and Introduction of the Railway Locomotive. By SAMUEL SMILES, Author of "Self-Help," &c. With Steel Portraits and numerous Illustrations. 8vo, Cloth, $3 00.

SMILES'S HISTORY OF THE HUGUENOTS. The Huguenots: their Settlements, Churches, and Industries in England and Ireland. By SAMUEL SMILES. With an Appendix relating to the Huguenots in America. Crown 8vo, Cloth, Beveled, $1 75.

SMILES'S SELF-HELP. Self-Help; with Illustrations of Character, Conduct, and Perseverance. By SAMUEL SMILES. New Edition, Revised and Enlarged. 12mo, Cloth.

SPEKE'S AFRICA. Journal of the Discovery of the Source of the Nile. By Captain JOHN HANNING SPEKE, Captain H. M. Indian Army, Fellow and Gold Medalist of the Royal Geographical Society, Hon. Corresponding Member and Gold Medalist of the French Geographical Society, &c. With Maps and Portraits and numerous Illustrations, chiefly from Drawings by Captain GRANT. 8vo, Cloth, uniform with Livingstone, Barth, Burton, &c., $4 00.

STRICKLAND'S (MISS) QUEENS OF SCOTLAND. Lives of the Queens of Scotland and English Princesses connected with the Regal Succession of Great Britain. By AGNES STRICKLAND. 8 vols., 12mo, Cloth, $12 00.

THE STUDENT'S HISTORIES.

France. Engravings. 12mo, Cloth, $2 00.
Gibbon. Engravings. 12mo, Cloth, $2 00.
Greece. Engravings. 12mo, Cloth, $2 00.
Hume. Engravings. 12mo, Cloth, $2 00.
Rome. By Liddell. Engravings. 12mo, Cloth, $2 00.
Old Testament History. Engravings. 12mo, Cloth, $2 00.
New Testament History. Engravings. 12mo, Cloth, $2 00.
Strickland's Queens of England. Abridged. Engravings. 12mo, Cloth, $2 00.

TENNYSON'S COMPLETE POEMS. The Complete Poems of Alfred Tennyson, Poet Laureate. With numerous Illustrations by Eminent Artists, and Three Characteristic Portraits. 8vo, Paper, 50 cts.; Cloth, $1 00.

THOMSON'S LAND AND THE BOOK. The Land and the Book; or, Biblical Illustrations drawn from the Manners and Customs, the Scenes and the Scenery of the Holy Land. By W. M. THOMSON, D.D., Twenty-five Years a Missionary of the A.B.C.F.M. in Syria and Palestine. With two elaborate Maps of Palestine, an accurate Plan of Jerusalem, and several hundred Engravings, representing the Scenery, Topography, and Productions of the Holy Land, and the Costumes, Manners, and Habits of the People. 2 large 12mo vols., Cloth, $5 00.

TICKNOR'S HISTORY OF SPANISH LITERATURE. With Criticisms on the particular Works, and Biographical Notices of Prominent Writers. 3 vols., 8vo, Cloth, $5 00.

VÁMBÉRY'S CENTRAL ASIA. Travels in Central Asia. Being the Account of a Journey from Teheran across the Turkoman Desert, on the Eastern Shore of the Caspian, to Khiva, Bokhara, and Samarcand, performed in the Year 1863. By ARMINIUS VÁMBÉRY, Member of the Hungarian Academy of Pesth, by whom he was sent on this Scientific Mission. With Map and Woodcuts. 8vo, Cloth, $1 50.

ENGLISHMAN'S GREEK CONCORDANCE. The Englishman's Greek Concordance of the New Testament: being an Attempt at a Verbal Connection between the Greek and the English Texts; including a Concordance to the Proper Names, with Indexes, Greek-English and English-Greek. 8vo, Cloth, $5 00.

WOOD'S HOMES WITHOUT HANDS. Homes Without Hands: being a Description of the Habitations of Animals, classed according to their Principle of Construction. By J. G. WOOD, M.A., F.L.S., Author of "Illustrated Natural History." With about 140 Illustrations, engraved by G. Pearson, from Original Designs made by F. W. Keyl and E. A. Smith, under the Author's Superintendence. 8vo, Cloth, Beveled Edges, $4 50.

WILKINSON'S ANCIENT EGYPTIANS. A Popular Account of their Manners and Customs, condensed from his larger Work, with some New Matter. Illustrated with 500 Woodcuts. 2 vols., 12mo, Cloth, $3 50.

MAURY'S (M. F.) PHYSICAL GEOGRAPHY OF THE SEA. The Physical Geography of the Sea, and its Meteorology. By M. F. MAURY, LL.D., late U.S.N. The Eighth Edition, Revised and greatly Enlarged. 8vo, Cloth, $4 00.

ANTHON'S SMITH'S DICTIONARY OF ANTIQUITIES. A Dictionary of Greek and Roman Antiquities. Edited by WILLIAM SMITH, LL.D., and Illustrated by numerous Engravings on Wood. Third American Edition, carefully Revised, and containing, also, numerous additional Articles relative to the Botany, Mineralogy, and Zoology of the Ancients. By CHARLES ANTHON, LL.D. Royal 8vo, Sheep extra, $6 00.

ANTHON'S CLASSICAL DICTIONARY. Containing an Account of the principal Proper Names mentioned in Ancient Authors, and intended to elucidate all the important Points connected with the Geography, History, Biography, Mythology, and Fine Arts of the Greeks and Romans; together with an Account of the Coins, Weights, and Measures of the Ancients, with Tabular Values of the same. Royal 8vo, Sheep extra, $6 00.

DWIGHT'S (REV. DR.) THEOLOGY. Theology Explained and Defended, in a Series of Sermons. By TIMOTHY DWIGHT, S.T.D., LL.D. With a Memoir of the Life of the Author. Portrait. 4 vols., 8vo, Cloth, $8 00.

FOWLER'S ENGLISH LANGUAGE. The English Language in its Elements and Forms. With a History of its Origin and Development, and a full Grammar. Designed for Use in Colleges and Schools. Revised and Enlarged. By WILLIAM C FOWLER, LL.D., late Professor in Amherst College. 8vo, Cloth, $2 50.

GIESELER'S ECCLESIASTICAL HISTORY. A Text-Book of Church History. By Dr. JOHN C. L. GIESELER. Translated from the Fourth Revised German Edition by SAMUEL DAVIDSON, LL.D., and Rev. JOHN WINSTANLEY HULL, M.A. A New American Edition, Revised and Edited by Rev. HENRY B. SMITH, D.D., Professor in the Union Theological Seminary, New York. Four Volumes ready. (*Vol. V. in Press.*) 8vo, Cloth, $2 25 per vol.

GODWIN'S (PARKE) HISTORY OF FRANCE. The History of France. From the Earliest Times to the French Revolution of 1789. By PARKE GODWIN. Vol. I. (Ancient Gaul). 8vo, Cloth, $3 00.

HALL'S (ROBERT) WORKS. The Complete Works of Robert Hall; with a brief Memoir of his Life, by Dr. GREGORY, and Observations on his Character as a Preacher, by Rev. JOHN FOSTER. Edited by OLINTHUS GREGORY, LL.D., and Rev. JOSEPH BELCHER. Portrait. 4 vols., 8vo, Cloth, $8 00.

HAMILTON'S (SIR WILLIAM) WORKS. Discussions on Philosophy and Literature, Education and University Reform. Chiefly from the *Edinburgh Review*. Corrected, Vindicated, and Enlarged, in Notes and Appendices. By Sir WILLIAM HAMILTON, Bart. With an Introductory Essay, by Rev. ROBERT TURNBULL, D.D. 8vo, Cloth, $3 00.

HUMBOLDT'S COSMOS. Cosmos: a Sketch of a Physical Description of the Universe. By ALEXANDER VON HUMBOLDT. Translated from the German, by E. C. OTTÉ. 5 vols., 12mo, Cloth, $6 25.

ROBINSON'S GREEK LEXICON OF THE TESTAMENT. A Greek and English Lexicon of the New Testament. By EDWARD ROBINSON, D.D., LL.D., late Professor of Biblical Literature in the Union Theological Seminary, New York. A New Edition, Revised, and in great part Rewritten. Royal 8vo, Cloth, $6 00.

www.ingramcontent.com/pod-product-compliance
Lightning Source LLC
LaVergne TN
LVHW021133110826
845150LV00005B/1018

* 9 7 8 1 4 2 5 5 4 8 6 3 6 *